3rd International Scientific and Practical Conference "Efficient Waste Treatment" (EWT 2021)

IOP Conference Series: Earth and Environmental Science
Volume 938

St. Petersburg, Russia
21 June 2021

ISBN: 978-1-7138-4920-9
ISSN: 1755-1307

Printed from e-media with permission by:

Curran Associates, Inc.
57 Morehouse Lane
Red Hook, NY 12571

Some format issues inherent in the e-media version may also appear in this print version.

Institute of Physics
Dirac House, Temple Back
Bristol BS1 6BE UK

Phone: 44 1 17 929 7481
Fax: 44 1 17 920 0979

techtracking@iop.org

Additional copies of this publication are available from:

Curran Associates, Inc.
57 Morehouse Lane
Red Hook, NY 12571 USA
Phone: 845-758-0400
Fax: 845-758-2633
Email: curran@proceedings.com
Web: www.proceedings.com

TABLE OF CONTENTS

Preface

Peer Review Declaration

Geotechnical Properties of Yellow Silty Sand Soil Modified by Bio-Stimulation with Presence of NiCl2 1
Jili Qu, Jinrui Cheng, Junfeng Wang

Resource-Saving Processing of Technological Solutions and Wastewater by Electrochemical Method (methods Overview) 9
N A Bykovsky, E A Kantor, M A Malkova, P A Rahman, L N Puchkova, N N Fanakova

Investigation of Dielectric Heating Method of Zinc-Containing Materials for Chlorides Removal 16
A G Ryazanov, G G Mikhailov, O V Khmeleva, Y D Savina, D M Galimov, A V Senin

The Vortex Layer Apparatus as a Source of Low-Grade Heat in the Process of Pretreatment of the Substrate Before Anaerobic Digestion 22
Andrey A Kovalev, Dmitriy A Kovalev, Victor S Grigoriev, Yuriy V Litti

Feed Additive for Quails Based on Fermented Brewer's Spent Grain 31
N.A. Yurina, N.D. Labutina, A.A. Danilova, A.B. Vlasov, B.V. Khorin, D.A. Yurin

Comparative Analysis of Methods for Determining the Flammability of Municipal Solid Waste as Railway Transport Cargo 35
L A Koroleva, A G Khaidarov, G K Ivakhnuk, T N Antoshina, A A Boyakhchyan

Engineering-Geological and Geoecological Aspects of Formation of Dry Dumps on Hydrodumps 42
Yu I Kutepov, N A Kutepova, Yu Yi Kutepov, A D Vasileva, A S Mukhina, R D Smirnov

Modeling of Chloroform Content in Drinking Water of Infiltration Water Intake by Periods of the Annual Cycle. Part 1. Extended Flood Period 50
M A Malkova, A I Vasileva, N D Minchenkov, E A Kantor

Metal Chips Preparation for Utilization Using Advanced Reagents 58
A Samukov, M Cherkasova, M Kuksov, S Dmitriev

Low-Waste Innovative Production Technology for Canned Quince Compote, Enriched with Natural Nutrients 66
M.M. Rakhmanova, M.E. Akhmedov, A.F. Demirova, Z.A. Abdulkhalikov, N.G. Zagirov

Asessment of the Impact of Vehicle Emissions on the Atmosphere of the City of Voronezh and Their Minimization 73
V. Ya. Manokhin, M. V. Manokhin

The Use of Web Technologies in the Transportation of Medical Waste Generated During the Treatment of COVID-19 on the Territory of the Russian Federation 86
V G Burlov, M V Mironova, V A Abramov, S Yu Kapitsyn

Geoecological and Economic Assessment of the Municipal Solid Waste Management System in the Kaluga Region 93
I N Lykov, E V Obolenskaya

The Role of Secondary Resource Market in the Development of the Extended Producer Responsibility Mechanism in Russia ... 100
M A Liubarskaia, N A Putinceva

Mining Wastes of the Russian North-West: Microbiological Prerequisites of Reclamation ... 107
E Abakumov, A Kimeklis, G Gladkov, E Andronov, A Zverev

Waste from Processing of Technical Grape Varieties in Poultry Nutrition ... 113
Raisat Akhmedkhanova, Haji Shabanov, Samira Aliyeva, Albina Alakayeva, Irina Musayeva, Patimat Hiramagomedova

Biohydrogen Production from Food Processing Wastewater by a Newly Isolated Thermophilic Bacterium ... 120
Yuriy V Litti, Elena A Zhuravleva, Andrey A Kovalev, Dmitriy A Kovalev, Inna V Katraeva, Sofiya N Parshina

Comparative Evaluation of Environmental Efficiency of the Municipal Solid Waste Treatment Scenarios for St. Petersburg Region with the Life Cycle Assessment Method ... 127
N V Mikhailova, A V Yasinskaya

Method for Milk Whey Microfiltration with Filtrate Pulsed Backpressure and Installation for Its Implementation ... 136
N N Korysheva, S A Titov, S V Shakhov, G S Tikhonov, I A Glotova, N A Galochkina

Combination of Ground Penetrating Radar and Seismic Surveys of the Tailings Dam ... 144
A Kalashnik, A Dyakov

An Environmental Engineering Approach to Wastewater Treatment with Bivalves in the Coastal Zone ... 150
AV Gudimov

Author Index

IOP Conf. Series: Earth and Environmental Science **938** (2021) 011001 doi:10.1088/1755-1315/938/1/011001

Dear colleagues,

I am glad to announce the release of the Efficient Waste Treatment conference proceedings (EWT-2021)!

Despite the pandemic conditions, we managed to successfully hold an online (zoom) conference and discuss possible solutions for waste management in terms of energy, environmental and economic factors. A competent approach allowed us not to postpone the event, which took place on June 21, 2021.

Conference model: Plenary session with keynotes (40 minutes for each keynote speech) + participants' presentations (20 minutes for each presentation).

There were two keynote speakers at the conference with extensive presentations:

- ✓ Svetlana Zueva, dr. eng., research scientists at the Department of Industrial Engineering, Information and Economy, University of L'Aquila (Italy).
- ✓ Natalia Politaeva, Doctor of Technical Sciences, Professor at the Graduate School of Hydraulic and Power Engineering Construction, Peter the Great St.Petersburg Polytechnic University (Russia).

Topics discussed were: Waste treatment in various regions and countries; Economic issues of waste management; Municipal solid waste; Water treatment and sewage sludge utilization; Industrial, construction and medical wastes; Mathematical modeling of ecological processes; Sustainable mining and processing of minerals; Agriculture & farming; Adaptive landscape farming systems; Environmental and construction engineering; Ecotourism.

Within the framework of the conference, a scientific committee worked, which included:

Michael Butusov, Professor, The University of Natural Resources and Life Sciences, Vienna, Austria

Aleksander Fedyukhin, Associate Professor, Moscow Power Engineering Institute, Moscow, Russia

Christoph Pfeifer, Professor, The University of Natural Resources and Life Sciences, Vienna, Austria

Natalia Politaeva, Professor, Peter the Great St.Petersburg Polytechnic University, St. Petersburg, Russia

Kerstin Kuchta, Professor, Hamburg University of Technology, Germany

Iryna Atamaniuk, Dipl.-Ing. Dr.techn., Hamburg University of Technology, Germany

Rafat Al Afif, Senior Scientist and lecturer, University of Natural Resources and Life Sciences, Vienna, Austria

Julia Smyatskaya, Associate Professor, Peter the Great St.Petersburg Polytechnic University, St. Petersburg, Russia

IOP Conf. Series: Earth and Environmental Science **938** (2021) 011001 doi:10.1088/1755-1315/938/1/011001

Svetlana Zueva, dr. eng., research scientists, Department of Industrial Engineering, Information and Economy, University of L'Aquila, Italy

Ida de Michelis, dr.eng., Department of Industrial Engineering, Information and Economy, University of L'Aquila, Italy

The best articles of the conference participants, which passed the peer review, were accepted for publication. We have done a great job and would like to share with you the main results.

We hope you find this article collection interesting and useful. Enjoy reading!

Prof. Politaeva Natalia,

Chairman of the EWT scientific committee

IOP Conf. Series: Earth and Environmental Science **938** (2021) 011002 doi:10.1088/1755-1315/938/1/011002

Peer review declaration

All papers published in this volume of EES have been peer reviewed through processes administered by the Editors. Reviews were conducted by expert referees to the professional and scientific standards expected of a proceedings journal published by IOP Publishing.

- **Type of peer review: Double-blind**
- **Conference submission management system: MeisterTask CRM**
- **Number of submissions received: 44**
- **Number of submissions sent for review: 36**
- **Number of submissions accepted: 22**
- **Acceptance Rate (Number of Submissions Accepted / Number of Submissions Received X 100): 50%**
- **Average number of reviews per paper: 2**
- **Total number of reviewers involved: 5**
- **Any additional info on review process: Each article was checked for scientific content, quality of the English language and technical formatting. In the absence of a scientific component of an article, authors right to revision was rejected. In other cases, correction notes were sent to authors.**
- **Contact person for queries: Anastasia Kulachinskaya, a.kulachinskaya@yandex.ru**

IOP Conf. Series: Earth and Environmental Science **938** (2021) 012001 doi:10.1088/1755-1315/938/1/012001

Geotechnical properties of yellow silty sand soil modified by bio-stimulation with presence of NiCl2

Jili Qu[1], Jinrui Cheng[1], and Junfeng Wang[1]

[1]School of Environment and Architecture, University of Shanghai for Science and Technology, Shanghai, 200093, China

Email: qujiliqwq@163.com

Abstract. Microbiologically induced calcite precipitation (MICP) is now widely tried to improve the quality of mostly sandy soil due to its application prospect to sustainability, environment-friendly and cost-efficiency. The present study is aiming to evaluate the feasibility of bio-stimulation, one of MICP techniques, for strengthening the engineering quality of Shanghai silty sand soil, a special type of soil located in low hillside of west Shanghai, as well as the influence of two factors: CaCl2 as a cementing material and NiCl2 as a co-factor, which was rarely researched before and can be viewed as the influence of urease activity on bio-stimulation result. The direct shear test, unconfined compressive test, supersonic wave test and calcite mass test have been carried out to evaluate engineering performance of yellow silty sand soil enhanced by bio-stimulation method as well as their relationship with concentration of CaCl2 and NiCl2 solutions. The result shows that bio-stimulation method can significantly promote the characteristics of Shanghai yellow silty sand soil and there seems to be extreme point for concentration of CaCl2 and NiCl2 solution in improving the quality of Shanghai yellow silty sand. Microstructure and composition of treated soil by bio-stimulation method have also been analyzed by Scanning Electron Microscope (SEM) and Energy dispersive Spectrometer (EDS), with giving the corresponding micro-mechanism of bio-cementation among soil particles.

1. Introduction

Microbiologically induced calcite precipitation (MICP) is a natural biogeochemical process in which calcite is precipitated within the soil environment due to microorganism activity [1-6]. MICP technique has been intensively researched in recent decades in such fields as geotechnical engineering etc for mending cracks, reducing permeation, improving strength and stiffness of sandy and silty soil, relieving liquefaction, eliminating dust etc. due to its property of sustainability and feasibility as well as environment-friendly [7]. The majority of traditional engineering improvement techniques on soil deplete too many resources in manufacturing processes and on-site operation, which sometimes even causes adverse impacts, such as toxic chemicals, massive carbon dioxide emissions etc., on the environment. It has been reported recently that the process of manufacturing the industrial concrete alone is responsible for about six percent of global CO_2 emission, which many experts think is the primary origin of social CO_2 discharge around the globe [8]. With a view to the ever-growing need for applications in geotechnics, as well as ecological, environmental and sustainable concerns, researchers and practitioners have been more and more apt to search for more advisable biogenic alternatives with minimal or no carbon footprint for the purpose of all kinds of engineering construction applications [9,10]. MICP is one of the prospective biogenic techniques used in recent decades in geotechnical circle for rock and soil improvement purpose. Ground improvement is an important branch of its applications, which has actually been widely researched especially for sandy soil and applied in the form of small number of experiments in recent decades, and has showed significant potential [11-15].

Shanghai yellow silty sand soil is a special type of soil, located in the west of Shanghai city, mainly distributed at the foots of low hills and nearby region, which is frequently encountered in geotechnical investigation and construction engineering projects. It is generally indispensable to improve the strength of the yellow silty sand soil before setting out to start a project. But the previous researches have not provided the related answers for improving the ground mainly composed of the local yellow silty sand soil with MICP technique, especially with bio-stimulation method. In addition, NiCl2, as a co-factor of urease has barely been researched before, which actually can be viewed as an important variable influencing the effectiveness in bio-stimulation process. Therefore, the purpose of the present study is to firstly evaluate the applicability of bio-stimulation technique to Shanghai yellow silty sand soil, and to secondly investigate the effect of concentrations of NiCl2 and CaCl2 on the solidification result for Shanghai yellow silty sand soil. The researches on above problems are expected to possess the global significance because the findings from the present study are likely to be used for reference in comparable regions with same or similar ground conditions across the world.

2. Materials and methods

2.1. Materials

2.1.1. Soil. The soil used in this study was collected from a construction site locating at the foots of Sheshan low hilly land in west of Songjiang district, Shanghai city. The particle size distribution was tested based on ASTM D422-63(2007) (ASTM 2007). The soil is classified as silty sand soil (SM) according to ASTM d2487 (ASTM 2011a). The soil properties and chemical composition are shown in Tables 1 and 2, respectively.

Table 1. Properties of silty sand soil from a construction site in Jinshan district

Property	Value
Soil classification	SM
D10	0.003mm
D30	0.018mm
D60	0.185mm
Coefficient of curvature	0.58
Coefficient of uniformity	61.67
Specific gravity (G_s)	2.69
d_{max} (e_{min})	1.95g.cm^{-3} (0.38)
d_{min} (e_{max})	1.54g.cm^{-3} (0.75)
Compacted soil density (Rd=38%)	1.66g.cm^{-3} (0.61)
Natural water content	3.0%

Table 2. Chemical composition of yellow silty sand soil at the foots of Sheshan low hilly land in the west of Songjiang district

Compound	SiO2	Al2O3	Fe2O3	FeO	TiO2	MgO	CaO	K2O	Na2O	P2O5	MnO
% by weight	69.56	14.48	1.11	2.57	0.43	0.44	2.19	5.25	3.72	0.1	0.1

The specific gravity of the sample was determined according to ASTM D854 (ASTM 2014), which was 2.69. The maximum dry unit weight of the silty sand soil was measured with vibrating table method according to ASTM D4253 (ASTM 2016a) and D4254 (ASTM 2016b). In the current research, a relative density of 38% as a medium density state and a moisture content of 15% were used in the test based on the previous research [16]. A moisture content of 15% was viewed as sufficient to help mixing the silty sand soil with the desired amounts of reactants and to achieve an appropriate moist condition for preparing the samples with the moist tamping method. A lower moisture content hindered the evenly and fully stirring of the silty sand soil with reactants, and a higher moisture content obstructed the thorough moist tamping of the mix of silty sand soil and reactants.

2.1.2. Reactants. The reactants used in the test can be divided into three types: enrichment solution (yeast extract), cementation solution ($CaCl_2$ and Urea) and co-factor of urease ($NiCl_2$). According to the previous research [16] that use varied enrichments with carbon and nitrogen sources to study the stabilization effect in stimulating the indigenous bacteria, yeast extract is the best enrichment.

Therefore the yeast extract is directly selected as the main enrichment material for culturing the urease-producing bacteria. The urea is the source of CO_3^{2-} due to that when it is hydrolyzed, the CO_3^{2-} can be generated, which can be combined with Ca^{2+} from $CaCl_2$ in water to form the calcite cementing the soil particles. However the concentration of $CaCl_2$ solution in intergranular void has a great effect on cementation effect of soil, so the effect of the amount of $CaCl_2$ on stabilization result is also one of the main focuses in this research, especially for Shanghai yellow silty sand soil. $NiCl_2$ is said to be an accelerator for enzymes produced by the bacteria. In this research, the function of $NiCl_2$ has also been focused in stimulating the activity of indigenous microorganism and urease it produces based on change of the solidification strength and stiffness.

2.2. Methods

2.2.1. Sterile treatment. In the present study, the function of $CaCl_2$ and $NiCl_2$ concentration as well as the culturing method in modifying the geotechnical properties of soil were studied. The amounts of the chemical compound were obtained from the studies of previous researchers that proved for sandy soil [4,16,17,18]. The detailed specification of the chemical solution applied in the experiment is shown in Table 3.

Because the present test involves the bacteria, therefore it is indispensable for all the apparatus and chemical solutions added to the silty sand soil to be sterile. The distilled water and solutions other than urea and yeast extract were sterilized by autoclave at 121°C for 15 min. Due to the decomposing of urea and yeast extract during autoclaving, therefore they were sterilized by needle filtration and then added aseptically to the solution.

All the samples for different type of test were prepared at a relative density of 38% and cultured in an environment with a room temperature of about 25°C, which is helpful to the thorough culturing and thriving in the activity of microorganisms and urease they produce.

2.2.2. Sample preparation. The silty sand soil obtained from the field firstly passes through 2mm sieve to remove the small stones, branches, leaves, etc. that could have been included in the silty sand soil in the course of forming history. Then the sample was mixed with the proper amount of solution to keep the water content of it at about 15%.

Samples prepared for unconfined compressive test (UCT) were densified in place in three layers into PVC spilt cylinders with 10 cm in height and 4 cm in inner diameter, respectively, so as to make the samples to reach the relative density of 38% after mixing the silty sand soil with appropriate amount of solution (enrichment solution and cementation solution, as presented in Table 3). All the equipment used in these tests was autoclaved immediately before tests began. All the samples were wrapped in plastic bag, placed into curing box that was kept at constant temperature of about 25°C and maintained in state of culturing 7 days before testing.

The UCT tests were conducted in the light of ASTM D2166M (ASTM 2016). The same test was repeated three times for reducing the random error in the test as much as possible and for the purpose of the reproducibility. The calcite content was considered as the same for the identical treatment regardless of direct shear test or unconfined compressive test. Therefore the increase rate of calcite content measured from direct shearing samples for same treatment was considered as the same as that from unconfined compressive samples.

For experiments of shear wave rate, the ultrasonic pulse velocity (UPV) experiments were carried out to appraise the shear stiffness of samples at small or micro strain. The instrument used is an ultrasonic pulse velocity tester (type PL-200) manufactured by Suzhou Tuoce instrument Co. Ltd. of Jiangsu province in P.R.China. As UPV test is a nondestructive experiment, samples could be able to undergo UPV experiment instantly prior to unconfined compressive test. UPV experiments were executed in line with ASTM D2845. The fundamentals of the UPV experiment is in accordance with the propagation theory of ultrasonic waves with frequency being higher than 20kHz, passing through a cylindrical specimen with a pair of sensing elements linked to its top and bottom using a specially made coupling gel. After ultrasonic waves sent from one end of the specimen are received, the time period from the moment they are produced to the moment they are received at another end of the specimen (termed the wave travelling time) is unscrambled by the UPV manufacturer's application software. The first marked trough is selected as the moment of ultrasonic wave arrival. The initial shear stiffness is able to be estimated from the following equation: $G_o = (\frac{d}{t_s})^2 \times \rho = V_s{}^2 \times \rho$, where G_0 is the initial stiffness of specimen; d is the spacing between the radio transmitter and receiver; t_s is shear-wave travelling time; ρ is the bulk unit mass of specimen; Vs is the shear-wave speed (d/ts). Shearing wave velocity experiment for any one treatment was executed for 3 times, with their average being taken to reduce the possible error as much as possible.

Table 3. Chemical recipe for making solutions required and sample designation

Group	Sample name	Solutions: Urea/$CO(NH_2)_2$ (mmol/L) (filtration)	Yeast (%) (filtration)	$CaCl_2$ (mmol/L) (autoclave)	$NiCl_2$ (mmol/L) (autoclave)
a	1	0	0	0	0
b	2	333	5	0	0
c	3	333	5	100	0
	4	333	5	200	0
	5	333	5	300	0
	6	333	5	400	0
d	7	333	5	0	0.0075
	8	333	5	0	0.01
	9	333	5	0	0.0125
	10	333	5	0	0.015
e	11	333	5	100	0.0075
	12	333	5	100	0.01
	13	333	5	100	0.0125
	14	333	5	100	0.015
f	15	333	5	200	0.0075
	16	333	5	200	0.01
	17	333	5	200	0.0125
	18	333	5	200	0.015
g	19	333	5	300	0.0075
	20	333	5	300	0.01
	21	333	5	300	0.0125
	22	333	5	300	0.015
h	23	333	5	400	0.0075
	24	333	5	400	0.01
	25	333	5	400	0.0125
	26	333	5	400	0.015

*Sterilization method: all chemicals were sterilized by autoclave except for urea that is sterilized by filtration. Curing method: all the samples were tested after 7 days curing under normal stresses of 50, 100, 150, 200kPa, respectively, which

corresponds to the normal stresses during the shear test later, respectively. Soil type used in the test is classified as SM (silty sand) according to ASTM d2487.

3. Results and discussions

3.1. Unconfined compressive test

In order to save space, the representative samples were selected to show the stress-strain relationship for unconfined compressive test. From figure 1, it can be observed that sample 21 has the highest unconfined compressive strength (UCS) of about 150 kPa, which increased by 275%. Among no 5, 13, 17, 21, 25, sample no 5 has the lowest increase of all samples treated with $NiCl_2$ and $CaCl_2$. This is maybe because it has been not treated by $NiCl_2$ as a co-factor of urease that can promote the activity of urease. Among No 1, 2, 9, No 2 and 9 have slight increase over control sample, this because No 2 obtained the urea and yeast that can provide enough CO_3^{2-}, which combine with indigenous Ca^{2+}, even though it did not obtain the external $CaCl_2$ as a Ca^{2+} source and $NiCl_2$ as a co-factor. Although No 9 has been treated with appropriate amount of $NiCl_2$, due to lack of enough external Ca^{2+} introduction, therefore its strength increase is also small. In addition we can also see that the failure strain for No 5, 13, 17, 21, 25 has been significantly increased compared to control sample. It has been increased by about 150% for No 5 compared to control sample. The above results are indicative of not only the significate increase in unconfined compressive strength but also in failure strain that increases the ductility of treated samples. The increase in unconfined compressive strength and strain failure are very important in project design, investigation, construction and maintenance of geotechnical engineering.

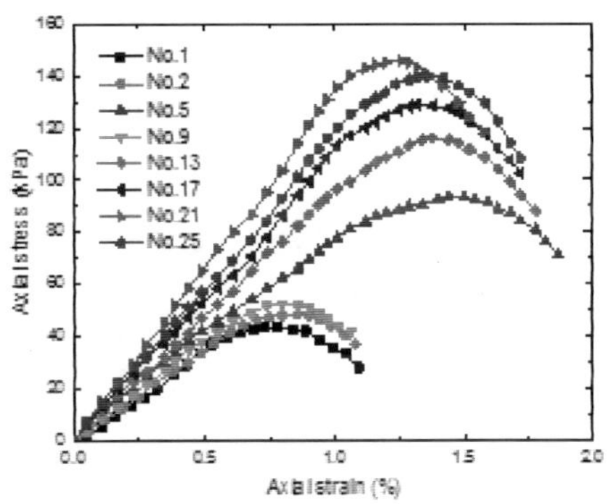

Figure 1. stress-strain relationship of representative sample under unconfined compressive test.

Figure 2 shows the unconfined compressive strength of all samples. The yellow bar is for control sample used for comparison. So it can be easily observed that for any content of calcite, 0,0125 mmol/L of NiCl2 is the most appropriate quantity for soil cementation even though the concentration of CaCl2 is 0 mmol/L. For 0.0125 mmol/L concentration of NiCl2 , when CaCl2 concentration is 300 mmol/L, the peak stress of unconfined compressive test is maximized. Therefore 0.0125 mol/L of NiCl2 concentration and 300 mmol/L of CaCl2 concentration are the most appropriate combination for cementation of shanghai yellow silty sand in environment used by the present test.

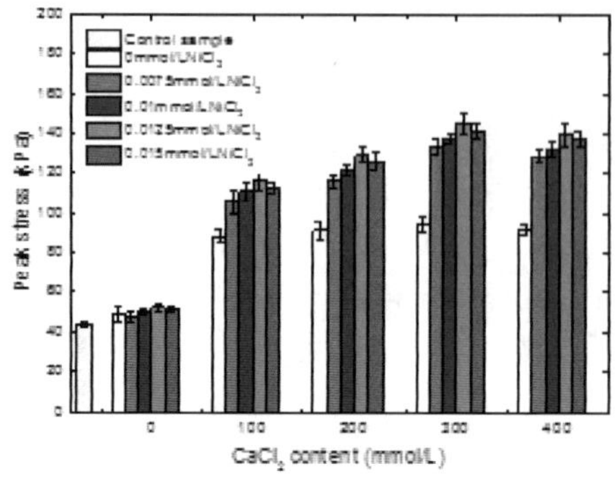

Figure 2. Histogram for unconfined compressive strength of all sample

3.2. Shear wave velocity.

Another important indicator for measuring the effect of solidification for soil is the change in stiffness of soil. Because there is a relatively fixed relationship between the sample stiffness and wave velocity that travels through the samples, the elastic wave speed traveling through the sample was used to represent the change in sample stiffness. Figure 3(a) shows the change of shear wave velocity over curing time for representative samples. From the change of the curve, it can be seen the shear wave velocity of sample No 5, 13, 17, 21, 25 has a considerable increase compared to control sample, which indicates the corresponding stiffness of above samples is also considerably increased, of which the shear wave velocity for no 21 is maximized. no 2 and no 9 had hardly been increased in shear wave velocity compared to control samples due to their insufficient treatment and cementation. In addition, from the above curve it can also be judged that cementation process is changed based on the time elapse. The higher the shear wave velocity, the higher the stiffness the sample, the better the cementation. Therefore, it can be deduced that the cementation process have almost been completed about 70% at the end of the first day, which is indictive of the bacteria activity and urease activity are the highest in the first day, than declining gradually from the second to seventh day. This is meaningful in practical engineering for process control. Figure 3 (b) shows histogram for the shear wave velocity of representative samples. It can be obviously seen that No 1, 2, 9 have the lowest wave velocity or the No 2, and 9 have the lowest increase in wave velocity over control sample.

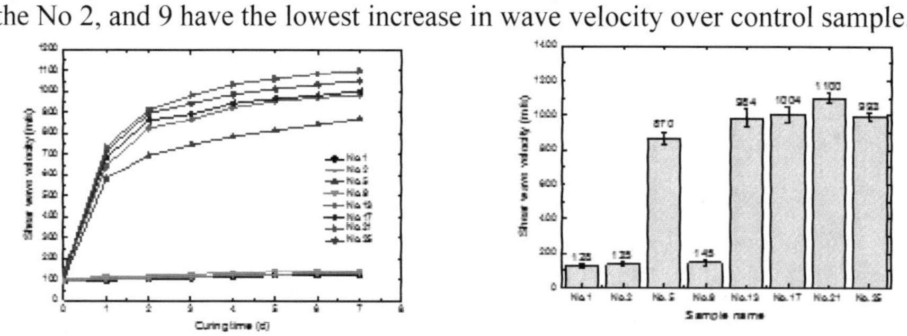

Figure 3. shear wave velocity change with curing time (a) and histogram for shear wave velocity of representative sample (b)

3.3. Calcite mass precipitated

Figure 4 (a) shows the calcite precipitated in representative samples. It can be seen that no 2 and no 9 has only very limited increase in calcite content compared to control sample, while no 25 has the highest calcite content that is 5.01%, which is increased by 1093% compared to control sample. The no 9 has only 1.73% increase in calcite content, which is increased by 307% compared to control sample. This just shows the difference between the presence and no presence of $NiCl_2$ as a co-factor. It can be concluded from the measuring of calcite content that the effect of solidification for soil with the present of $NiCl_2$ is increased by almost 256%. This is just the $NiCl_2$ that make a difference, which is indicative of the importance of the co-factor in solidification of soil. In order to compared the effect of $NiCl_2$ and $CaCl_2$ on the precipitation of calcite, the 3D multiple regression diagram was constructed as shown in figure 4 (b). it can be easily observed that the calcite content is generally increased with the concentration of $CaCl_2$ solution, and maximized at about 300 mmol/L of $CaCl_2$ and when concentration of $NiCl_2$ is approximately 0.0125 mmol/L the content of calcite is maximized. It is worth noting that in Figure 4 (a) the No 25 has the highest content of calcite but it don't have the highest strength in unconfined compressive test. This may be due to that the excess calcite is precipitated on the surface of soil particle that is not functioned as cementation, while the calcite precipitated at the contact point among soil particles can function as cementing role promoting strength.

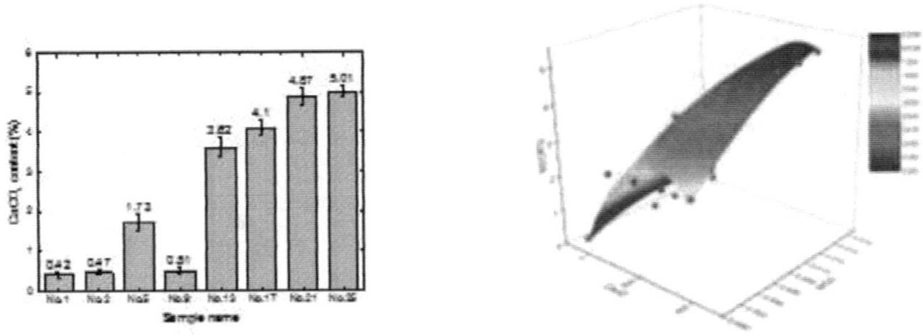

Figure 4. Calcite mass precipitated in representative samples (a), and 3D multiple regression diagram for the effect of concentration of $NiCl_2$ and $CaCl_2$ on the precipitation of calcite (b)

4. Conclusion

The present study aims to detect the applicability and feasibility of using the bio-stimulation technology to strengthen the Shanghai yellow salty sand, as well as effect of $NiCl_2$ and $CaCl_2$ on the solidification result. The related macro and micro scale test have been carry out . the conclusion drawn from above analysis are as follows:

1. The shanghai yellow silty sand is appropriate for being densification by bio-stimulation technology;
2. The concentration of $NiCL_2$ and $CaCl_2$ has significant influence on the consolidation effect of Shanghai yellow silty sand based on bio-stimulation;
3. 0.0125 mmol/L of $NiCl_2$ and 300 mmol/L $CaCl_2$ solution may be the best combination for solidifying Shanghai yellow silty sand, which can make the shear strength and unconfined compressive strength increased by at most 59%, and 275%, respectively;
4. The calcite content precipitated cannot be directly used to measure the strength of treated soil, instead, the calcite content precipitated at the contact points between particles may be the best indicator for measuring the soil strength. Sometimes the excess calcite may be precipitated on the surface of the particle that is not very helpful to promote the soil strength.

The feasibility of bio-stimulation for improving the geotechnical property of Shanghai yellow silty sand located in the hillside and foot areas of Sheshan hill in Songjiang district is significantly important for geotechnical investigation, design, construction etc in future projects, due to its characteristics of sustainability, environment-friendly and no or little CO_2 emission, which is completely in agreement with the development trend across the modern world.

References

[1] DeJong J T, Fritzges M B and Nüsslein K 2006 Microbially induced cementation to control sand response to undrained shear *Journal of Geotechnical and Geoenvironmental Engineering* **132(11)** 1381–1392

[2] Whiffin V S, Van Paassen L A and Harkes M P 2007 Microbial carbonate precipitation as a soil improvement technique *Geomicrobiology Journal* **24(5)** 417–423

[3] Van Paassen L A 2009 *Biogrout, ground improvement by microbial induced carbonate precipitation*: PhD Thesis (Delft University of Technology)

[4] Mortensen B M, Haber M J, Dejong J T and et al 2011 Effect of environmental factors on microbial induced calcium carbonate precipitation *Journal of Applied Microbiology* **111(2)** 338–349

[5] Dejong J T, Soga K, Kavazanjian E and et al 2013 Biogeochemical processes and geotechnical applications: progress, opportunities and challenges *Geotechnique* **63(4)** 287–301

[6] Dejong J T, Gomez M G, Waller J T and et al 2017 Influence of Bio-Cementation on the Shearing Behavior of Sand Using X-Ray Computed Tomography *Geotechnical Frontiers* 871–880

[7] Yu T, Souli H, Péchaud Y and Fleureau J-M 2020 Optimizing protocols for microbial induced calcite precipitation (MICP) for soil improvement – a review *European Journal of Environmental and Civil Engineering* 1–16

[8] Achal V and Mukherjee A 2015 A review of microbial precipitation for sustainable construction *Construction and Building Materials* **93** 1224–1235

[9] Ashraf M S, Azahar S B and Yusof N Z 2017 Soil improvement using MICP and biopolymers: A review *IOP Conference Series: Materials Science and Engineering* **226(1)**

[10] Chang I, Im J and Cho G C 2016 Introduction of microbial biopolymers in soil treatment for future environmentally-friendly and sustainable geotechnical engineering *Sustainability* **8(3)** 251

[11] Sun X, Miao L, Tong T and Wang C 2018 Improvement of microbial-induced calcium carbonate precipitation technology for sand solidification *Journal of Materials in Civil Engineering* **30(11)** 04018301-1~8

[12] Dakhane A, Das S, Hansen H and et al 2018 Crack healing in cementitious mortars using enzyme-induced carbonate precipitation: Quantification based on fracture response *Journal of Materials in Civil Engineering* **30(4)** 04018035-1~10

[13] Eujine G N, Chandrakaran S and Sankar N 2017 Accelerated subgrade stabilization using enzymatic lime technique *Journal of Materials in Civil Engineering* **29(9)** 04017085-1~7

[14] Sravan M V and Nagaraj H B 2017 Potential use of enzymes in the preparation of compressed stabilized earth blocks *Journal of Materials in Civil Engineering* **29(9)** 04017103-1~8

[15] Bundur Z B, Bae S, Kirisits M J and et al 2017 Biomineralization in self-healing cement-based materials: investigating the temporal evolution of microbial metabolic state and material porosity *Journal of Materials in Civil Engineering* **29(8)** 04017079-1~8

[16] Kiasari M A, Pakbaz M S and Ghezelbash G R 2019 Comparison of effects of different nutrients on stimulating indigenous soil bacteria for biocementation *Journal of Materials in Civil Engineering* **31(6)** 04019067-1~13

[17] Gat D, Tsesarsky M, Wahanon A and et al 2014 Ureolysis and MICP with model and native bacteria: implications for treatment strategies *Geo-Congress 2014 Technical Papers* 1713–1720

[18] Gat D, Ronen Z and Tsesarsky M 2016 Soil bacteria population dynamics following stimulation for ureolytic microbial-induced $CaCO_3$ precipitation *Environmental Science & Technology* **50(2)** 616–624

Resource-saving processing of technological solutions and wastewater by electrochemical method (methods overview)

N A Bykovsky[1], E A Kantor[1], M A Malkova[1], P A Rahman[1], L N Puchkova[1], and N N Fanakova[1]

[1]Ufa State Petroleum Technological University, 1, Kosmonavtov Street, Ufa, 450062, Bashkortostan, Russian Federation

E-mail: nbikovsky@list.ru

Abstract. The paper summarizes the data on the wastewater formation from a number of plants of chemical, engineering, energy and leather industries and its processing in electrolyzers of various designs to obtain the substances that can be used in various production processes. In the ethylenediamine production, wastewater occurs at the stages of amines rectification and of ethylenediamine dihydrochloride neutralization. Recycling of effluents generated during the amine rectification in a two-chamber electrolysis cell allows the ethylenediamine and ammonia to be returned to the production process. The processing of ethylenediamine dihydrochloride solution in the membrane electrolyzers saves alkali for the neutralization process. Electrochemical processing of distilled liquid in soda production makes possible to obtain calcium hydroxide, sodium hydroxide and hydrochloric acid or chlorine, which can be used in various industrial processes. Wastewater generated during the demineralized water production contains a significant amounts of sodium sulfates and chlorides. The processing of these effluents in a membrane electrolyzer yields sulfuric acid and alkali, which are suitable for the regeneration of ion-exchange filters. After processing in an electrodialysis apparatus the wastewater generated during the chromium and nickel coatings, is concentrated to values that allow them to be used to prepare an electrolyte adjustment. The leather industry effluents contains an amount of trivalent chromium and sulphides. Electrochemical processing of these effluents leads to almost complete extraction of chromium and sulfides.

1. Introduction

The plants of chemical, engineering, energy and leather industries are concentrated in many regions of Russia. As a result of their activities, a significant amount of wastewater is formed. The paper summarizes the data on the wastewater formation and its processing by the electrochemical method in order not only to purify, but also to obtain a number of elements that can be used in production. The results presented in this paper can be useful for enterprises that have a technology similar to the mentioned.

In the ethylenediamine production, based on the interaction of 1,2-dichloroethane with ammonia, the ethylenediamine dihydrochloride is formed as an intermediate compound. To release the ethylenediamine, the dihydrochloride is decomposed by treatment with alkali, and then evaporated and rectificated [1, 2]. As a result of evaporation obtained ammonia is sent to the stage of amination of 1,2-dichloroethane, amines are sent to rectification to release amines, and the saturated sodium chloride solution containing 1-2% of amines. During the rectification process, the aqueous fraction containing about 1% of ethylenediamine and about 2% of ammonia is formed in the upper part of the

IOP Conf. Series: Earth and Environmental Science **938** (2021) 012002 doi:10.1088/1755-1315/938/1/012002

rectification column. Thus, the ethylenediamine production is followed by wastewater formation, which is the saturated solution of sodium chloride with an admixture of amines and the rectification aqueous fraction containing ethylenediamine and ammonia. Per 1 ton of ethylenediamine it is formed about 3 m^3 of saturated sodium chloride solution and about 14 m^3 of the rectification water fraction. Since the content of amines and ammonia in these effluents significantly exceeds the maximum allowable concentrations, they are subjected to multiple dilutions.

In the production of soda ash by the ammonia method [3, 4], after the release of ammonia and carbon dioxide, a formed distilled liquid contains about 100 g / l of $CaCl_2$, about 50 g / l of NaCl and about 2 g / l of other impurities. Per each ton of produced soda after ammonia regeneration it is formed up to 9 m^3 of distilled liquid. At present, the distilled liquid is almost not processed and from the distillation station is first sent to the storage ponds (the so-called white seas), and then discharged into water bodies, that leads to its pollution and salinity. After a special treatment it is possible to use the distilled liquid for injection into oil wells in order to maintain reservoir pressure. However, this method of its utilization is possible when the production of soda ash is located in the area of oil production [3]. The method of distilled liquid treatment to obtain $CaCl_2$ and NaCl is well-known [3]. This method is based on the three-stage evaporation process, which causes the multi-stage production, the complex technological equipment, the use of gas from lime kilns and the high energy consumption. So to treat 1 m^3 of distilled liquid according to this technology it is needed 13.2 m^3 of gas from lime kilns and 2.31 GJ of energy.

The plants of many industries consume demineralized water. Such water, in particular, is used to feed the boilers of thermal power plants. To obtain the demineralized water the ion-exchange technology is used [5, 6]. The process of demineralized water production takes place in two stages. The first stage (pre-treatment) includes the river water liming and it filtering through the mechanical filters. At this stage up to 65% of Ca^{2+} and Mg^{2+} ions, up to 70% of Fe^{2+} ions and up to 80% of suspended solids are extracted. The second stage is the water treatment in ion exchange filters. Ion exchange treatment is realized in a two-stage scheme. The first stage filters are loaded with a KU-2-8 cation exchanger and an AN-31 low-base anion exchanger. In filters of the second stage, a cation exchanger of the KU-2-8 type and a high-base anion exchanger of the AV-17 type are used. The regeneration of cation-exchange and anion-exchange filters is carried out with 4% solutions of sulfuric acid and sodium hydroxide, respectively. After regeneration, the filters are washed with demineralized water. The regeneration and washing of ion-exchange filters results in appearing of some solutions which salts concentration significantly exceeds its MAC.

Electrochemical metal coatings for various products is widely used in different product areas: mechanics engineering, automotive, radio-electronic industry, production of precision instruments, and so on. About 70% of engineering plants have galvanic shops or sections [7, 8]. As a result of the electroplating operations it appear wastewaters containing non-ferrous metals and other chemicals.

In the tanning industry, at the stages of soaking, washing and ashing, the waste waters contain up to 3 g / l of sulphides. In the process of leather tanning, the generated waste waters contain 2.9 ÷ 3.2 g / l of trivalent chromium and have 3.2 ÷ 3.8 of pH. The analysis of the resulting wastewater shows that before discharge, it require the purification or multiple dilutions.

It should be noted that these waste waters contain significant amounts of various salts. The purpose of this paper is to study the process of electrochemical effluents processing, that will allow not only to purify waste water, but also to obtain substances that can be used in the production process [9-13].

2. Material and research methods

The study of the wastewater treatment was carried out in electrolyzers with ion-exchange membranes, schematically shown in the figure 1.

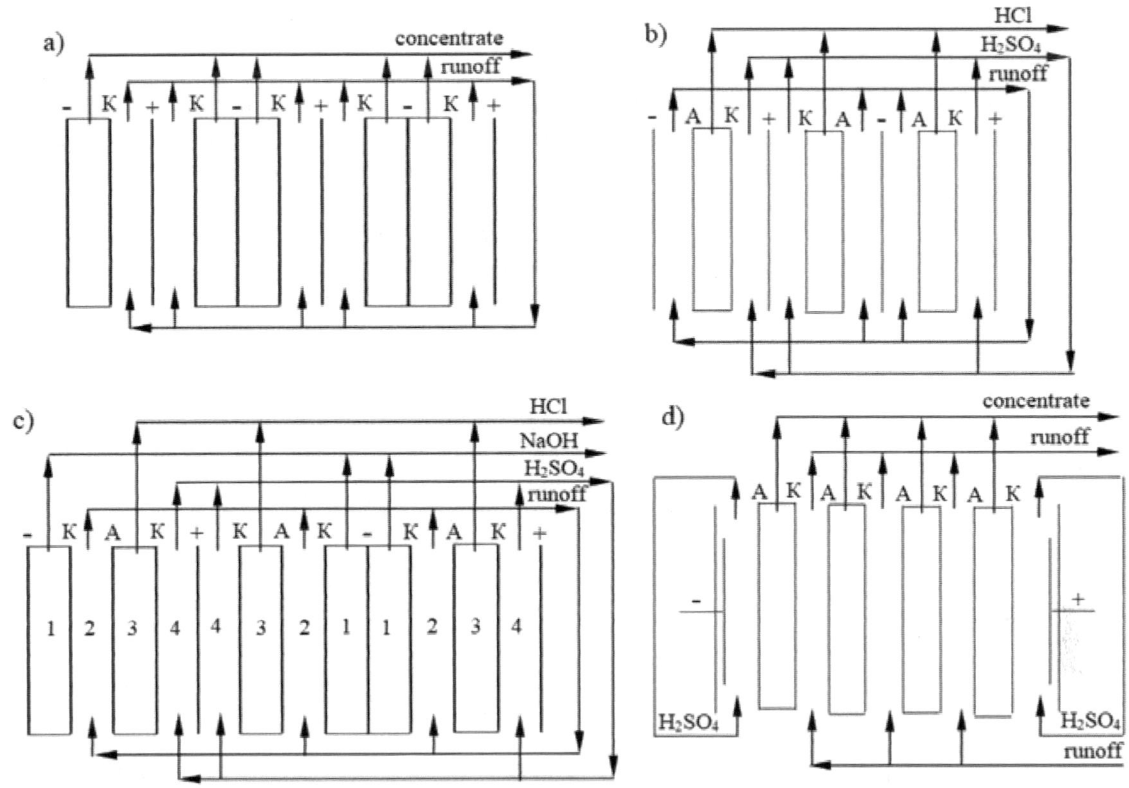

\- - cathode, + - anode, K - cation exchange membrane, A - anion exchange membrane

Figure 1. Electrolysis plant diagrams: a) two-chamber, b) three-chamber, c) four-chamber, d) electrodialysis

Electrolyzers represent a filter-press assembly of identical cells separated by electrode plates. Stainless steel plates served as cathodes, and titanium plates with ruthenium oxide coat served as anodes. Each cell consisted of a) 2 chambers; b) 3 cameras; c) 4 chambers separated by MK-40 cation-exchange membranes and MA-40 anion-exchange membranes. The cameras are the 1.9 mm thick PVC frames. To prevent contact of the membranes with the electrodes, a perforated stretched mesh of calendered vinyl plastic was placed in each chamber. The work surface of each ion-exchange membrane was 50 cm^2. Waste water circulated in the desalination chambers. The solutions generated in the concentration chambers left its as it accumulated. The experiments time was determined by reaching the maximum concentration of solutions generated in the concentration chambers.

3. Experimental results and discussion

3.1. Ethylenediamine production

The processing of waste water generated on the upper rectifying plates was studied [14]. The studies were carried out in a two-chamber electrolyzer with a cation-exchange membrane. The waste water circulated in the anode chambers, formed on the upper rectifying plates. The concentrate was generated in cathode chambers. It is shown that during the electrochemical processing it is possible to obtain a concentrate containing up to 117 g / l of ethylenediamine and up to 107 g / l of ammonia. At the same time, the concentration of ethylenediamine and ammonia in the treated effluent is at the maximum allowed level. The volume of the concentrate is more than 10 times less than the volume of the processed effluent. This will allow to use it in the ethylenediamine production to obtain ammonia solution.

To exclude alkali from the ethylenediamine dihydrochloride processing, the process of electrochemical processing of a dihydrochloride solution was studied. The electrochemical processing can be organized in two-, three- and four-chamber electrolyzers with ion-exchange membranes. It is possible to obtain chlorine in a two-chamber electrolyzer, hydrochloric acid in a three-chamber electrolyzer, and ethylenediamine and hydrochloric acid in a four-chamber electrolyzer. The ethylenediamine dihydrochloride processing in two- and three-chamber electrolysers allows to reduce the alkali consumption in dihydrochloride decomposition, and the ethylenediamine dihydrochloride processing in a four-chamber electrolyzer makes it possible to completely exclude alkali from the process.

3.2. Soda ash production
The authors studied the distiller liquid processing to obtain $Ca(OH)_2$, NaOH, HCl and Cl_2. The process consisted of two stages [15]. At the first stage, the distilled liquid was treated with sodium hydroxide. As a result, a calcium hydroxide precipitate was formed, which was filtered off. The resulting filtrate contained approximately 170 g / l of NaCl and 0.49 g / l of $CaCl_2$. At the second stage, the filtrate was processed in two-, three- and four-chamber electrolyzers. Depending on the apparatus design, it is possible to obtain alkali with a concentration of up to 450 g / l, hydrochloric acid with a concentration of up to 145 g / l and chlorine. The processing 1 m^3 of distiller liquid allows to obtain 87.4 kg of $Ca(OH)_2$, 43.9 kg of NaOH, 5.8 kg of HCl and 124 kg of Cl_2. The resulting products are offered for use as raw materials and semi-finished products at the plants of the Sterlitamak industrial hub.

3.3. Demineralized water obtaining
The composition of wastewater generated in the regeneration and washing of anion-exchange filters was investigated [16], also as these waters processing in membrane electrolyzers to obtain sodium hydroxide, sulfuric acid and chlorine. The averaged impurities concentrations in the effluent obtained during the anion-exchange filters regeneration are presented in Table 1.

Table 1. Averaged impurities concentration in the effluent of anion-exchange filters

Name	The content of the effluent		MPC, mg / dm^3
	First stage	Second stage	
Na^+, mg / dm^3	3840	8380	200
Cl^-, mg / dm^3	1020	670	350
SO_4^{2-}, mg / dm^3	1140	4310	500
OH^-, mg / dm^3	1680	4950	pH=6÷9
SiO_3^{2-}, mg / dm^3	0,135	879	10

The impurities concentration in the waste water, resulting from the regeneration and washing of anion-exchange filters, significantly exceeds the maximum allowable one. However, the presented results indicate that the wastewater formed during the regeneration and washing of anion-exchange resins of water treatment plants can be used to obtain alkalis and acids by treating its in a membrane electrolyzer.

The content of some elements in waste water generated during the cation-exchange filters regeneration exceeds the value of the maximum allowable concentrations for these elements, that is shown in Table 2.

Table 2. Exceeding the maximum allowable concentration

Name	Exceeding the MPC, times	
	First stage	Second stage
Sodium	4.9	–
Calcium	3.3	–
Magnesium	3.5	–
Potassium	4.45	2.56
Iron	9.6	15.2
Titanium	6.3	2.2
Nickel	11.5	11
Chromium	1.12	–
Lead	14.3	4
Cadmium	1.5	

A significant excess of MPC is noted for iron (15.2 times), nickel (11.5 times) and lead (14.3 times).

Waste water from the cation exchange filters regeneration is not suitable for electrochemical processing. This is due to the fact that during the electrolysis process in the cathode chamber, precipitates of calcium, magnesium and other metals hydroxides will fall out.

The data analysis in the study of the composition of wastewater generated in the regeneration and washing of ion-exchange filters indicates that the effluents of anion-exchange filters can be used as a starting product for the electrochemical acid and alkali production. The wastewater subjected to electrochemical treatment contains the same substances, only in different quantities. The treatment of waste water generated during the regeneration and washing of anion-exchange filters was investigated. The current efficiency, calculated on the results of impurities extraction from the effluents, decreases with increasing current density for the waste water of both stages. In this case, the values of the current output in the second stage are higher than in the first stage. The maximum current efficiency 92.5% for the effluent of the first stage and 98.6% for the effluent of the second stage corresponds to a current density of 400 A / m^2.

The concentration of sodium hydroxide is determined by the operating mode of the electrolyzer. The maximum NaOH generation rate for the initial moment of time gradually decreases, tending to a constant value. The sodium hydroxide content in the cathode chamber increases with increasing of the current density. In this case, the concentration of NaOH obtained from the waste water from the second stage is higher than the first. So after 8 hours of electrolysis at a current density of 1200 A / m^2, sodium hydroxide with a concentration of 196.3 g / dm^3 is obtained from the effluent of the first stage, and sodium hydroxide with a concentration of 232.2 g / dm^3 is obtained from the effluent of the second stage. The current efficiency calculated from NaOH generated in the apparatus cathode chamber agree closely with the current efficiency calculated on the impurities extraction results.

The maximum concentration of H_2SO_4 is noted at the current density of 1200 A / m^2 and makes 129.2 g / dm^3 in the first stage and 84.9 g / dm^3 in the second stage. Cl^- ions migrating into the anode chamber are not completely discharged at the anode. As a result, the hydrochloric acid is present in the anode chamber. No dependence of the hydrochloric acid concentration from the time of waste water electrochemical treatment of both stages was found. The average value of the hydrochloric acid concentration in the anode chamber is 2.45 g / dm^3 for the first stage and 0.57 g / dm^3 for the second stage. Experiments carried out with industrial waste water showed the generation of silicic acid in addition to sulfuric and hydrochloric acid in the anode chamber of the electrolyzer. The concentration of silicic acid is almost $5.8 \cdot 10^3$ times lower than the concentration of sulfuric acid.

Thus, the electrochemical processing of waste water generated during the regeneration and washing of anion-exchange filters, both of the first and second stages of a water treatment plant, makes it possible not only to impurities extraction from the effluent, but also to obtain sulfuric acid and sodium hydroxide, suitable for use in the regeneration of ion-exchange filters.

3.4. Chrome plating and nickel plating

Analysis of electroplating sites waste water shows that the waste water generated in the process of chrome plating and nickel plating of parts contains up to 64 mg / l of chromium and 32 mg / l of nickel. When replacing the electrolyte in the baths its content increases to 294 mg / l and up to 260 mg / l, respectively.

The electrodialysis extraction of chromium and nickel from wastewater generated in galvanic production was studied. It is shown that the concentration of chromium and nickel in the effluent can be reduced to 5 mg / l. After dilution in the all-plant effluents, the concentration of both chromium and nickel becomes below the maximum allowable. The consumption of electrical power for the purification process at the maximum reached extraction degree makes 1.58 kW·h / m^3 and 3.75 k·Wh / m^3 when nickel is removed from chromium. Along with the wastewater purification from chromium and nickel electrolytes, the electrodialysis method makes it possible to obtain concentrates of these electrolytes containing up to 1100 mg / l of chromium and 1230 mg / l of nickel. At the same time, the volume of the concentrate is more than 50 times less than the volume of the processed wastewater. This phenomenon allows to use the chromium and nickel concentrates to prepare solutions used to adjust the electrolyte in galvanic baths. Thus, it is possible to create a closed cycle of water use in galvanic production.

3.5. Leather production

To purify wastewater from chromium (III), it is proposed to treat it in the cathode chamber of a diaphragm electrolyzer. Water decomposition occurs at the cathode:

$$2H_2O - 2e \rightarrow H_2\uparrow + OH^-,$$

as a result, hydroxyl ions enter the electrolyte. Two processes are possible at the anode:

$$2Cl + 2e \rightarrow Cl_2\uparrow,$$

$$2H_2O + 2e \rightarrow O_2\uparrow + 4H^+,$$

in this case, hydrogen ions enter the electrolyte. The diaphragm separating the cathode and anode spaces of the electrolyzer complicates the process of hydrogen and hydroxyl ions recombination. As a result, the catholyte pH increases, which leads to the $Cr(OH)_3$ precipitate formation. The experimental results obtained in the study of the chrome planting process show that in this way it is possible to extract chromium from the effluent almost up to 100%. After a simple treatment, the chromium hydroxide precipitate can be used to prepare a tanning agent or used in metallurgical production.

The study of wastewater purification from sulfides was carried out by its sedimentation as iron sulfide, which can be obtained by sulfide effluents treatment in an electrolyzer with a soluble iron anode. In this situation, scrap metal can be used, and the resulting semi-finished product of iron sulfide with scrap metal can be sent for processing to metallurgical plants. The maximum sulfide recovery is reached at an anode current density of 133 A / m^2 and makes 99.2%.

4. Conclusion

In conclusion, it should be noted that the electrochemical processing of technological solutions and wastewater from a number of plants in Sterlitamak, and therefore other similar enterprises, allows not only to purify wastewater, but also to obtain products that are applicable in production. At the same time, no reagents are used during the processing.

Acknowledgements

The work was performed within the framework of the Ministry of Science and Higher Education of the Russian Federation state task in the field of scientific activity, publication number FEUR - 2020 -

0004 «Solving urgent problems and researching processes in petrochemical industries accompanied by flows of multiphase media»

References

[1] Lebedev N N 2016 *Chemistry and technology of basic organic and petrochemical synthesis: educational for universities* (Moscow: Alliance)
[2] Zagtdullin R N 2004 RU Patent No 2, 226, 189 (27 March 2004)
[3] Tkach G A, Shaporev V P and Titov V M 1998 *Production of soda using low-waste technology* (Kharkov: KhGPU)
[4] Shatov A A 2017 Soda ash production – from past to new technologies *Scientific review. Basic and applied research* **1** Available at: http://www.scientificreview.ru/ru/article/view?id=8 (accessed: 08.05.2020)
[5] Haddad M, Bazinet L and Barbeau B 2019 Eco-efficient treatment of ion exchange spent brine via electrodialysis to recover NaCl and minimize waste disposal *Science of The Total Environment* **690** 400–409
[6] Jiang S, Sun H, Wang H, Ladewig B P and Yao Z 2021 A comprehensive review on the synthesis and applications of ion exchange membranes *Chemosphere* **282** 130817
[7] Vinokurov E G, Arsenkin K V, Grigorovich K V and Bondar V V 2006 Electrodeposition of chromium coatings modified with dispersed particles *Metal protection* **42(3)** 312–316
[8] Walsh F C and Leon C P 2014 Versatile electrochemical coatings and surface layers from aqueous methanesulfonic acid *Surf. Coat. Tech.* **259** 676–697
[9] Kingsbury R S, Zhu S, Flotron S and Coronell O 2018 Microstructure Determines Water and Salt Permeation in Commercial Ion-Exchange Membranes *ACS Appl Mater Interfaces* **10(46)** 39745–39756
[10] Hyder A G, Morales B A, Cappelle M A, Percival S J, Small L J, Spoerke E D, Rempe S B and Walker W S 2021 Evaluation of Electrodialysis Desalination Performance of Novel Bioinspired and Conventional Ion Exchange Membranes with Sodium Chloride Feed Solutions *Membranes* **11(3)** 217
[11] Raka Y D, Bock R, Karoliussen H, Wilhelmsen O and Stokke Burheim O 2021 The Influence of Concentration and Temperature on the Membrane Resistance of Ion Exchange Membranes and the Levelised Cost of Hydrogen from Reverse Electrodialysis with Ammonium Bicarbonate *Membranes* **11(2)** 135
[12] Stenina I, Golubenko D, Nikonenko V and Yaroslavtsev A 2020 Selectivity of Transport Processes in Ion-Exchange Membranes: Relationship with the Structure and Methods for Its Improvement *Int. J. Mol. Sci.* **21(15)** 5517
[13] Hassanvand A, Wei K, Talebi S, Chen G Q and Kentish S E 2017 The Role of Ion Exchange Membranes in Membrane Capacitive Deionisation *Membranes* **7(3)** 54
[14] Bykovsky N A, Puchkova L N, Fanakova N N, Khalimov A G and Makarenko O A 2019 Recycling the Wastewater in the Production of Ethylenediamine *Chemical and Petroleum Engineering* **54** 787–794
[15] Bykovsky N A, Rahman P A, Puchkova L N and Fanakova N N 2017 Electrochemical Recycling of Still Waste Liquid in Ammonia Soda Production *Key Engineering Materials* **743** 342–346
[16] Bykovsky N A, Kantor E A, Rahman P A, Puchkova L N and Fanakova N N 2018 Composition of wastewater formed during regeneration of anion-exchange filters of water treatment plants *IOP Conference Series: Earth and Environmental Science* **194(9)** 092003

IOP Conf. Series: Earth and Environmental Science **938** (2021) 012003 doi:10.1088/1755-1315/938/1/012003

Investigation of dielectric heating method of zinc-containing materials for chlorides removal

A G Ryazanov[1], G G Mikhailov[1], O V Khmeleva[2], Y D Savina[1], D M Galimov[1], and A V Senin[1]

[1]South Ural State University, 76, Lenin Prospekt, Chelyabinsk, 454080, Russia.
[2]Far Eastern State Technical Fisheries University, 52b, Lugovaya Str., Vladivostok, 690087, Russia.

E-mail: ryazanov_andrei@mail.ru

Abstract. The technological development of the world community makes it necessary to resolve the issues of the produced industrial residues utilization. Worn-out galvanized metal cars, ceilings and other metal structures require orderly utilization. Utilization of galvanized scrap may be carried out in Electric Arc Furnaces (EAF) to obtain intermediate steel and EAF dust. Electric Arc Furnaces dus contains zinc, so it is used as a secondary raw material for the production of metallic zinc. Waelz oxides are formed in Waelz kilns, after the primary processing of EAF dust. Waelz oxide contains halides that must be removed first. Halides may be removed by heating to temperatures of about 1000°C. Dielectric (microwave) heating is a promising and environmentally friendly method for material processing. Microwave heating is carried out without burning natural gas, which leads to decrease of waste gases volume. The work experimentally confirmed the possibility of zinc-containing materials heating at electromagnetic emission exposure. The duration of products heating up to the temperature of 1000°C was 128 - 188 s. The residual content of chloride ions in the calcined products is less than 0.05 wt%.

1. Introduction

Metallic zinc and its alloys are mainly used for surface galvanizing. Galvanized products protect the material from premature corrosion. Galvanized scrap is mainly used in the automotive and construction industries [1-2]. Technological development leads to a gradual increase in the volume of zinc consumption, on average by 1.1-1.5%, including an increase in the consumption of galvanized products. As a result of its use, galvanized scrap is sent for processing to steel enterprises.

Galvanized scrap is mainly processed in Electric Arc Furnaces to produce intermediate steel and dust. Remelting one ton of steel forms about 15-25 kg of Electric Arc Furnaces (EAF) dust [3]. The chemical composition of EAF dust depends on many factors including the original composition of the processed scrap [4-6]. Typical zinc content in Electric Arc Furnace dust is 17-25 wt%. [7-8]. The disposal of EAF dust requires specialized landfills to reduce the negative environmental impact. The disposal of EAF dust is the most important task. Solving the problem of processing EAF dust will reduce the negative impact on the environment. EAF dust processing may be beneficial in recovering metallic zinc [9]. The Waelz process is considered to be a traditional method of Electric Arc Furnaces (EAF) dust processing [10-11]. Electric Arc Furnaces dust may be processed together with zinc residue of hydrometallurgical zinc production [12-13]. The product of zinc-containing materials

IOP Conf. Series: Earth and Environmental Science **938** (2021) 012003 doi:10.1088/1755-1315/938/1/012003

processing is Waelz oxide. Waelz oxide contains a significant amount of chlorides and fluorides (halides), which must be removed before loading into sulfuric acid solutions.

The technology of Waelz oxide washing in aqueous solution or in aqueous alkaline solution is used for halogens removal [13-16]. The washing method is based on the interaction of metal chlorides and fluorides with an alkali to transfer halogens into solution. In this case, the metals mainly remain in the target product - washed Waelz oxide. Washing technology requires the utilization of significant amount of waste water. The alternative method for halogens removal is the technology of pyrometallurgical calcination of Waelz oxide. The method of halogens removal from Waelz oxide is based on the difference between the evaporation temperatures of metal chlorides and fluorides compared to the evaporation temperatures of zinc and lead oxides. The calcination temperature is maintained at a level of 750 - 1100°C and created due to the combustion of natural gas [17-18].

Microwave heating is a promising and ecological method for heat treatment of Waelz oxide [19]. Microwave heating is widely used in the food, rubber and textile industries as well as in the processes of wood and carving wood drying [20-21]. The studies of using microwave heating in metallurgical processes are carried out, such as a recovery of metals from pyrolusite and Electric Arc Furnaces dust, and a number of hydrometallurgical leaching processes [22-25].

This work is aimed at increasing the volume of processing of secondary zinc-containing raw materials and, as a result, reducing the volume of negative environmental impact on nature. The aim of this work is to study the method of dielectric (microwave) heating of zinc-containing materials, which removes chlorides.

2. Materials and methods

Reagents zinc chloride $ZnCl_2$, $PbCl_2$ and zinc oxide ZnO were used as the initial material. The chloride-ion content was from 0.65 to 1.37 wt%. The chemical composition of the calcine products was determined by X-ray spectral microanalysis on a Jeol JSM-7001F Electron Scanning Microscope with an EDS Oxford INCA X-max 80 energy dispersive attachment.

The heating of the zinc-containing product by ultrahigh-frequency electromagnetic radiation was carried out in a modernized Samsung microwave furnace, with a power consumption of 1400 W, an operating frequency of 2450 MHz and a chamber volume of 28 liters. To capture gaseous and dusty fumes, consisting of chlorides and oxides, the furnace was equipped with a cyclone and a sanitary water scrubber. To store heat energy inside the furnace, the chamber volume is lined with fiberglass. A thermocouple of the "TXA" type was used to control the temperature of the sample during heating. Heating of zinc-containing products was carried out in a corundum crucible located in the lower part in the center of the chamber. The layout of sample in the microwave furnace is shown in Figure 1.

HSC Chemistry software package was used for thermodynamic modeling of phase and chemical transformations of Waelz oxide components upon heating in an oxidizing environment.

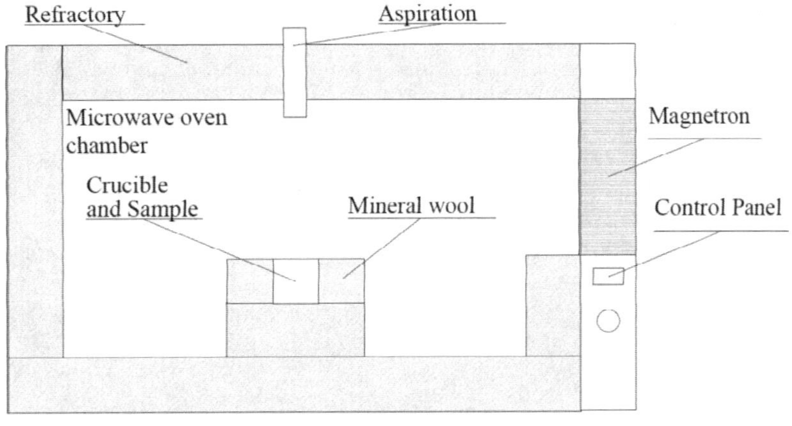

Figure 1. Sample layout.

3. Results and Discussion

3.1. Thermodynamic analysis of chemical and phase transformations

The purpose of thermodynamic analysis is to study changes in the composition of the solid and gaseous phases as a result of heating a model mixture of ZnO-$ZnCl_2$-$PbCl_2$-PbO substances. The initial data of the system for calculating the thermodynamic model are shown in Table 1.

Table 1. Initial data for thermodynamic calculation.

№	Material (species)	Content		Mass fraction, wt.%
		mole	kg	
1.	Solid, including:	**8160**	**69.4**	**100**
1.1.	ZnO (zinc oxide)	791.5	64.4	92.86
1.2.	$ZnCl_2$ (zinc chloride)	6.5	0.9	1.28
1.3.	PbO (lead oxide)	17.3	3.9	5.58
1.4.	$PbCl_2$ (lead chloride)	0.7	0.2	0.29
2.	Gas, including:	**9375**	**264.8**	**100**
2.1.	$N_2 + O_2$ (air)	8928.6	257.6	97.3
2.2.	CH_4 (natural gas)	401.8	6.4	2.4
2.3.	H_2O (steam)	44.6	0.8	0.3

The calculation results are shown in graphical form in Figure 2 and Figure 3. The content of dominant substances, such as ZnO in the condensed part of the system and N_2, O_2, CH_4, H_2, H_2O, CO and CO_2 in the gas phase, are not conventionally shown in the graphs. It can be seen that, even at room temperature, lead chloride is more stable than zinc chloride.

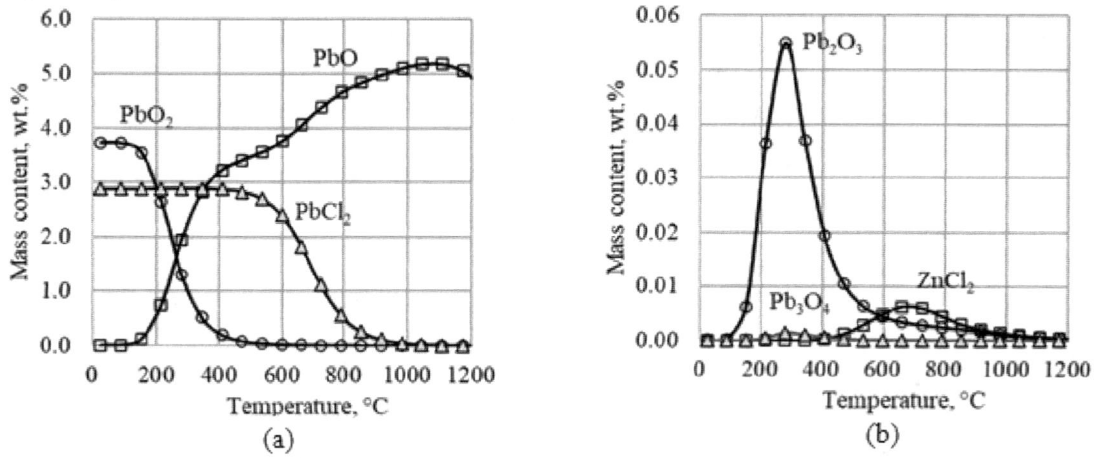

Figure 2. Effect of temperature on the mass of substances in the solid phase.

In the solid phase, the main component is zinc oxide ZnO (not shown in figure 3), its amount does not change significantly. Chlorides of zinc and lead, when heated, gradually pass into the gas phase. Intensive evaporation of lead chloride begins in the temperature range 500 - 600°C, zinc chloride in the temperature range 700 - 800 °C. On the graphs, these processes manifest themselves as a decrease in the mass content of the corresponding substances in the condensed phase and an increase in their content in the gas phase. At temperature of 1045 °C, the residual content of lead (II) chloride and zinc (II) chloride in the condensed part of the system is 0.03 and 0.001 wt%, respectively.

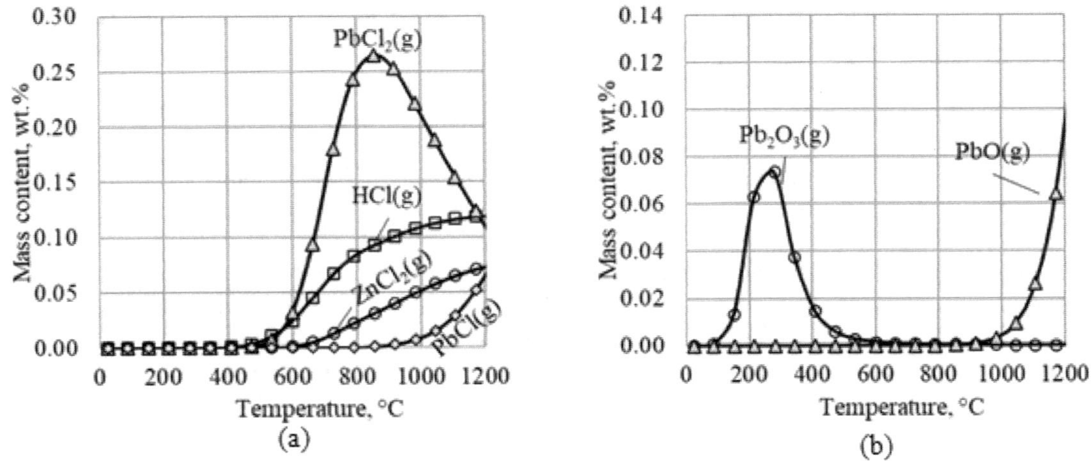

Figure 3. Effect of temperature on the mass of substances in the gas phase.

The main components of the gas phase are nitrogen N2, oxygen O2, carbon dioxide CO2 and water vapor H2O, formed from the combustion of natural gas. The vapor interacts with chlorides to form hydrogen chloride HCl in the gas phase and metal oxides. At temperatures above 1000°C, lead (II) oxide begins to pass into the gas phase.

It follows from the calculations that as a result of calcination, the chloride ion content in the solid product may be reduced up to 0.01 wt%.

3.2. Influence of microwave electromagnetic field on heating of samples.

During heating by microwave electromagnetic radiation, the temperature of zinc-containing mixtures samples was monitored using a TXA thermocouple and a high-resistance electronic multimeter.

The temperature was measured in two samples differing in the content of zinc chloride: sample No. 1 - 0.99 wt.% $ZnCl_2$; sample No. 2 - 2.38 wt.% $ZnCl_2$. The concentration of lead chloride is 0.53 wt% in both samples, the rest is zinc oxide. The time dependences of the heating temperature of the samples are presented in Figure 4.

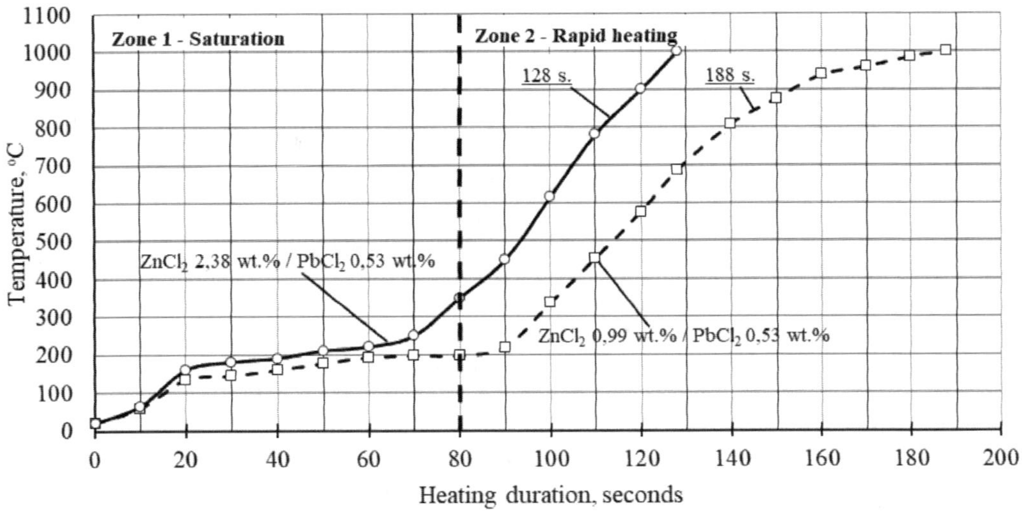

Figure 4. Sample temperature as a function of heating duration by the microwave electromagnetic field.

The graph is divided into two main areas: zone 1 - saturation, there is an accumulation of electromagnetic radiation in the samples; zone 2 - accelerated heating, the sample heats up sharply. Temperature dependences have the same character for different samples. In this case, the beginning of intensive heating of sample No. 1 with a lower content of zinc chloride begins 30 seconds later than sample No. 2 with a high content of zinc chloride. In the interval of intense temperature increase from 300 to 800 °C, the difference between the heating lines of samples No. 1 and No. 2 remains at the level of 30 seconds.

Based on the results of temperature control, the average heating rate of samples of zinc-containing mixtures to a temperature of 1000°C was calculated: sample No. 1 - 5.3 °C/s; sample No. 2 - 7.8 °C/s. Sample No. 2 with a high content of zinc chloride has a higher heating rate. Zinc chloride has a high degree of conversion of the energy of the microwave electromagnetic wave into thermal energy, therefore, the more zinc chloride in the mixture, the more intensively the sample heats up.

According to the results of chemical analysis, the concentration of chloride ion in the calcined samples was less than 0.05 wt%.

4. Conclusion

Thermodynamic analysis of changes in the composition of zinc-containing model mixtures of the ZnO – $ZnCl_2$ – $PbCl_2$ – PbO system upon heating to 1300 °C was performed using the HSC Chemistry software package. It has been established that intense evaporation of lead chloride begins in the temperature range 500 - 60°C, zinc chloride in the temperature range 700 - 800°C. At temperatures above 1000°C, the content of chloride ion in the condensed calcined products may be reduced to 0.01 wt%.

The processes of heating by electromagnetic radiation of samples of zinc-containing mixtures of the ZnO – $ZnCl_2$, $PbCl_2$ system, containing 0.99 and 2.53 wt% $ZnCl_2$, have been investigated in laboratory conditions. The average heating rate was 5.3 and 7.8°C/s, respectively.

It was found that the residual chloride ion content in the calcined products was less than 0.05 wt%.at calcination duration of 600 seconds.

Acknowledgements

The reported study was funded by RFBR, project number 19-38-90260.

References

[1] Bian J, Zhu Y, Liu X H and Wang G D 2006 Development of hot dip galvanized steel strip and its application in automobile industry *Journal of Iron and Steel Research International* **13(3)** 47–50

[2] Yeomans S R 2004 Galvanized steel in concrete: an overview *Galvanized Steel Reinforcement in Concrete* 1–289

[3] Guézennec A G, Huber J C, Patisson F, Sessiecq P, Birat J P and Ablitzer D 2005 Dust formation in electric arc furnace: birth of the particles Powder technology **157(1-3)** 2–11

[4] Alpatova A A, Simonyan L M, and Isakova N S 2016 Dust formation in the arc heating of zinc-plated steel *Steel in Translation* **46(5)** 303–308

[5] Simonyan L M, Alpatova A A and Demidova N V 2019 The EAF dust chemical and phase composition research techniques *Journal of Materials Research and Technology* **8(2)** 1601–1607

[6] Alpatova A A, Simonyan L M and Isakova N S 2016 Dust formation in the arc heating of zinc-plated steel *Steel in Translation* **46(5)** 303–308

[7] Piret N L 2012 EAF dust processing: rotary hearth a potential substitute for the Waelz Kiln? *World Metall Erzmetall* **65(5)** 306

[8] Gamroth M and Mager K 2010 SDHL waelz technology: state of the art for recycling of zinccontaining residues *Proceedings LEAD-ZINC '2010*, TMS, Vacouver, Kanada, 861 p

[9] Nemchinova N V, Chernykh V E, Tyutrin A A and Patrushov A E 2016 Extraction of zinc and iron from electrosmelting dust *Steel in Translation* **46(5)** 368–372
[10] Wu C C, Chang F C, Chen W S, Tsai M S and Wang Y N 2014 Reduction behavior of zinc ferrite in EAF-dust recycling with CO gas as a reducing agent *Journal of environmental management* **143** 208–213
[11] Suetens T, Klaasen B, Van Acker K and Blanpain B 2014 Comparison of electric arc furnace dust treatment technologies using exergy efficiency *Journal of cleaner production* **65** 152–167
[12] Akhtamov F E and Nishonov B U 2016 On the question of processing zinc cakes *Theory and technology of metallurgical production* **1(18)**
[13] Stoychev S, Minchev E, Kyurkchiev A and Radonov G 2020 Technologies for Treatment of Zinc-Containing Waste from Metallurgy in KCM AD *9th International Symposium on Lead and Zinc Processing* (pp 799–809) Springer, Cham
[14] Sar S, Sundqvist Öqvist L, Sparrman T, Engström F and Samuelsson C 2019 Characterization of Double Leached Waelz Oxide for Identification of Fluoride Mineral *Metals* 9**(3)** 361
[15] Antuñano N, Cambra J F, and Arias P L 2016 Fluoride removal from Double Leached Waelz Oxide leach solutions as alternative feeds to Zinc Calcine leaching liquors in the electrolytic zinc production process *Hydrometallurgy* **161** 65–70
[16] Güresin N and Topkaya Y A 1998 Dechlorination of a zinc dross *Hydrometallurgy* **49(1–2)** 179–187
[17] Panshin A M, Kozlov P A, Zatonsky A V, Gizatulin O V and Ivakin D A 2010 Development of a technology for calcining Waelz oxide using a large-sized tube furnace *Non-ferrous metals* **(5)** 45–48
[18] Sahin F, Derin C and Yucel O 2000 Chloride removal from zinc ash
[19] Ryazanov A G, Senin A V and Galimov D M 2021 Purification of Zinc Oxide from Chlorides Using Microwave Radiation *Key Engineering Materials* **887** 172–177. Trans Tech Publications Ltd
[20] Jones D A, Lelyveld T P, Mavrofidis S D, Kingman S W and Miles N J 2002 Microwave heating applications in environmental engineering – a review *Resources conservation and recycling* **34(2)** 75–90
[21] Das S, Mukhopadhyay A K, Datta S and Basu D 2009 Prospects of microwave processing: An overview *Bulletin of materials science* **32(1)** 1–13
[22] Agrawal S, Rayapudi V and Dhawan N 2018 Microwave reduction of red mud for recovery of iron values *Journal of sustainable metallurgy* **4(4)** 427–436
[23] Ye L, Peng Z, Wang L, Anzulevich A, Bychkov I, Tang H, Jiang T and et al 2018 Preparation of core-shell iron ore-biochar composite pellets for microwave reduction *Powder Technology* **338** 365–375
[24] Anzulevich A P, Kalganov D A, Anzulevich S N, Bychkov I V and Peng Z March 2020 Optimal microwave heating of biochar containing iron ore pellets *Journal of Physics: Conference Series* **1461(1)** 012007
[25] Wang L, Peng Z, Lin X, Ye Q, Ye L, Zhang J, Jiang T and et al 2021 Microwave-intensified treatment of low-zinc EAF dust: A route toward high-grade metallized product with a focus on multiple elements *Powder Technology* **383** 509–521

IOP Conf. Series: Earth and Environmental Science **938** (2021) 012004 doi:10.1088/1755-1315/938/1/012004

The vortex layer apparatus as a source of low-grade heat in the process of pretreatment of the substrate before anaerobic digestion

Andrey A Kovalev[1], Dmitriy A Kovalev[1], Victor S Grigoriev[1], and Yuriy V Litti[2]

[1]Federal Scientific Agroengineering Center VIM
[2]Federal Research Centre "Fundamentals of Biotechnology" of the Russian Academy of Sciences

E-mail: kovalev_ana@mail.ru

Abstract. The limiting stage of anaerobic digestion (AD) of organic waste is the hydrolysis of particulate organic matter. One of the most promising and energy-efficient methods of waste pretreatment before AD is its processing in a vortex layer apparatus (VLA). The work was aimed at experimentally determining the energy characteristics of VLA. The description of the experimental plant and its energy balance were given. The ranges of electricity consumption and heat generation were shown. According to experimental data, the coefficient of conversion of electricity into heat was in the range of 0.367-0.515 at a cooling water temperature of 10.6-11.7 oC. Based on the experimental data obtained, it can be argued that VLA is a stable source of low-grade heat that can be utilized using heat pumps.

1. Introduction

Organic wastes contain both readily degradable compounds, from which biogas is formed during one-stage anaerobic digestion, and difficult-to-decompose compounds, which require a longer retention time in the reactor for their conversion into biogas. The limiting stage of anaerobic digestion (AD) is the hydrolysis of the particulate organic matter. Soluble organic compounds, which can be further converted into biogas, are formed during hydrolysis. Therefore, the yield of biogas during AD is directly dependent on the ability of waste to biodegrade and, accordingly, on the rate of hydrolysis. One of the technological methods for increasing the bioavailability of organic waste is their pretreatment before fermentation in digesters. [1]

Proper preparation of waste is important in AD technology. The main tasks of preparation are:

1. separation of coarse inclusions (stones, boards, large debris, etc.);
2. homogenization of the mixture;
3. heating the mixture to the process temperature.

For this, various equipment is used, such as sieves, centrifuges, grinders, homogenizers, heat exchangers, etc. This equipment has significant energy consumption, therefore, the search and development of new technical solutions for the preparation of organic waste for AD is an urgent scientific and engineering task.

One of the most promising and energy-efficient methods of preparing a substrate for AD is processing it in a vortex layer apparatus (VLA). The vortex layer in VLA is created by the action of a rotating magnetic field [2]. VLA was shown earlier to be widely used in many industries, from the processing of mineral raw materials to the chemical, construction, food, cosmetic and pharmaceutical industries, for grinding raw materials [3-5]. One of the actual ways of using VLA is the processing of household and agricultural waste [6-7]. Also, the positive effect of pretreatment of various organic substrates in VLA on the characteristics of AD, in particular on the kinetics of methanogenesis, volatile solids (VS) removal efficiency, the content of methane in biogas, and parasites removal was shown [8-11]. However, in the recent literature, knowledge on the use of VLA in AD systems is scarce and value can be added.

Previous studies have shown that the use of VLA for a complex pretreatment of the initial substrate and digestate in the AD system will provide a synergistic effect of the following electrophysical impacts [8-11]:

- mechanical grinding;
- biochemical impact by the introduction of ferromagnetic particles into the pretreated substrate;
- electromagnetic impact.

However, these studies were carried out with the same operating parameters of VLA, namely:

- constant frequency of the electromagnetic field (50 Hz);
- constant amount of working bodies (0.4 kg of ferromagnetic ferromagnetic needles per 1 liter of the working chamber). Only the hydraulic retention time (HRT) in the working chamber VLA was changed.

For wider adoption of VLA into the AD system, it is necessary to determine the energy characteristics of VLA under various modes of its operation. The data obtained should be used in the development of the energy balance of the AD system. In addition, biogas heat pumps have a huge potential to be worldwide used for heating and cooling purposes.[12]

The purpose of this work is to experimentally determine the energy characteristics of the vortex layer apparatus under various operating modes as a source of low-grade heat for the subsequent use of the data obtained in the development of the energy balance of the technological line for anaerobic digestion of organic waste.

2. Materials and methods

2.1. Experimental plant

To study the energetic characteristics of VLA in the laboratory of bioenergetic and supercritical technologies, an experimental plant was developed for substrate pretreatment. The technological scheme of the experimental plant is shown in Figure 1, and the general view of the experimental plant is shown in Figure 2.

The installation includes the following elements:

The VLA working chamber (1) is located inside the VLA coil (3), equipped with a cooling jacket. During the operation of VLA, a significant part of the consumed electricity is converted into heat.

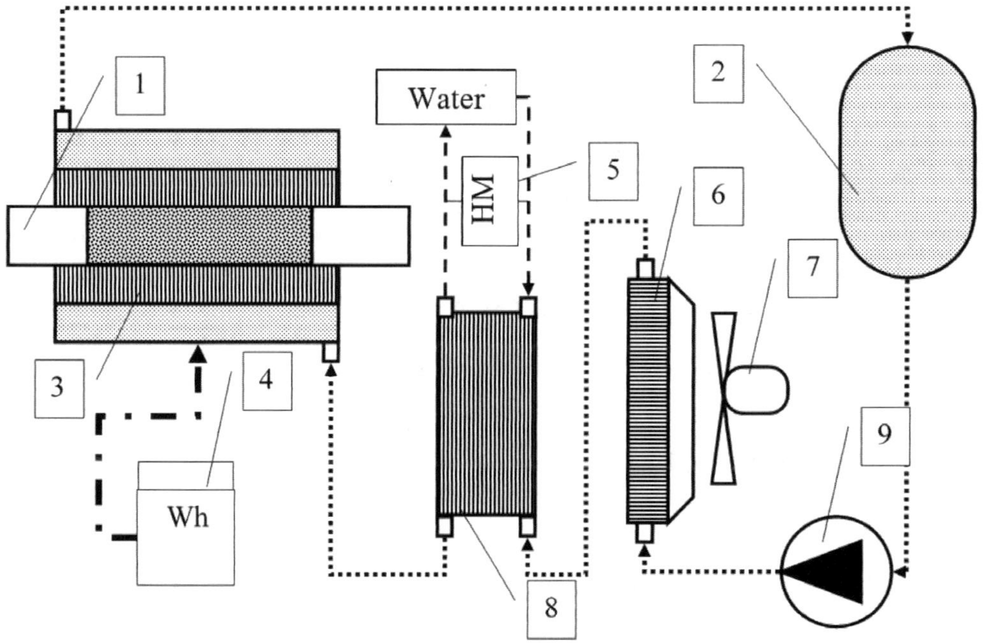

Figure 1. Technological diagram of the experimental setup (1 – VLA working chamber; 2 - expansion tank of transformer oil; 3 - VLA coil; 4 - electricity meter; 5 - heat meter; 6 - oil-air radiator; 7 - fan; 8 - plate heat exchanger (oil-water) 9 - coolant (transformer oil) circulation pump).

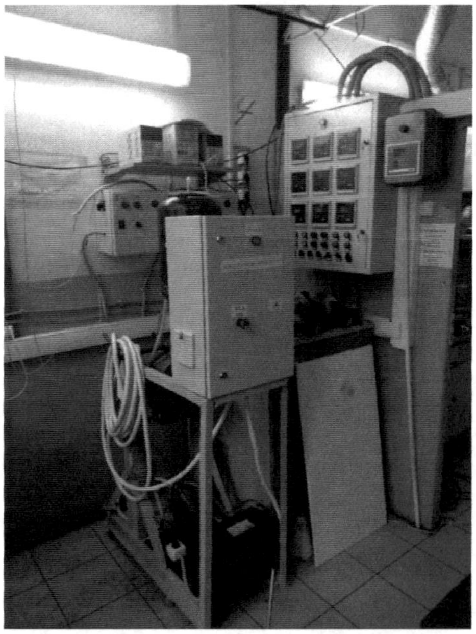

Figure 2. General view of the experimental plant.

To maintain the operating temperature of the VLA coil (3), a cooling circuit is provided, which includes an expansion tank (2) in series, heat exchangers (6) (oil-air) and (8) (oil-water) and a coolant circulation pump (9). Transformer oil of the GK brand is used as a heat carrier (coolant). The cooling circuit works as follows: the coolant circulation pump (9) supplies transformer oil to the water/oil-type plate heat exchanger (8), from where the cooled oil enters the cooling jacket of the VLA coil (3). From the cooling jacket of the VLA coil (3), the oil enters the expansion tank (2) and then into the oil-air radiator (6), which is equipped with a fan (7). From the oil-air cooler (6), the coolant circulation pump (9) feeds the transformer oil further along the cooling circuit. The oil-water plate heat exchanger (8) uses tap water to cool the oil.

A brazed plate heat exchanger GPLK 20-10 (Russia) was used as an oil-water heat exchanger.

A three-phase meter NIK 2303 ARP1 (Ukraine) was used as an electricity meter.

An ELF-M heat meter (Russia) with resistance thermometers of the Pt500 type as temperature sensors was used as a heat energy meter.

2.2. Energy balance

The general view of the energy balance of the experimental setup is as follows:

$$E_P + E_{VLA} = Q^{VLA} + N^{VLA} \quad (1)$$

where E_P is the specific amount of energy spent on the drive of the equipment of the cooling circuit of the VLA, $(kWh)^{electric\ power}/kg_{init}$; where kg_{init} is a kilogram of VS in the initial organic waste;

E_{VLA} - the specific amount of energy spent on the operation of the VLA, $(kWh)^{electric\ power}/kg_{init}$;

Q^{VLA} - the specific amount of heat given off by the VLA coil, $(kWh)^{heat\ power}/kg_{init}$;

N^{VLA} - the energy spent on the course of a set of processes (movement of working bodies, magnetostriction, electrolysis, etc.) in the VLA working chamber, $(kWh)^{electric\ power}/kg_{init}$.

In this case, only a part of the electricity supplied to the VLA is converted into thermal energy:

$$Q^{VLA} = k^Q \cdot E^{VLA} \quad (2)$$

where k^Q is the coefficient of conversion of electricity into heat on the VLA coil.

2.3. Setting up an experiment

The experimental setup worked without the processed liquid in two frequency ranges of the electromagnetic field:

- reduced frequency mode (**LF**) – VLA operation in the range of 41-49 Hz;
- operating frequency mode (**OF**) - VLA operation in the range of 45-80 Hz.

These ranges were chosen due to the results of a preliminary experiment, according to which, at a frequency of 40 Hz, the operation of the VLA is disrupted (the ferromagnetic needles cease to rotate). At a frequency above 80 Hz, due to the small amplitude of rotation of the ferromagnetic needles, the processing efficiency in VLA is significantly reduced.

The frequency step in the **LF** mode was 1 Hz, and in the **OF** mode - 5 Hz.

In both modes, the power consumption in kWh was determined.

In the **OF** mode, the following parameters were also determined:

- temperature of cooling water at the inlet and outlet of the heat exchanger, °C;
- cooling water flow, l / min.

The duration of the VLA operation was 10 minutes at each frequency. At the same time, the starting parameters were determined after 30 minutes of VLA being idle with the cooling turned on, and the operating parameters were determined immediately after 10 minutes of operation at the previous frequency.

The initial data for the OF mode are given in table 1.

Table 1. Initial data for the OF mode.

The initial temperature of cooling water, °C	Frequency, Hz							
	45	50	55	60	65	70	75	80
Starting	11	10,6	11	10,6	11,3	11,3	11,4	11,7
Operating	11,6	11	10,6	11	10,9	11	11,4	11,5

3. Results and discussion

3.1. Reduced frequency mode

For the LF mode, the power consumption was determined in the frequency range 41-49 Hz with a step of 1 Hz. The dependence of the starting and operating power consumption on the frequency of the electromagnetic field is shown in Figure 3.

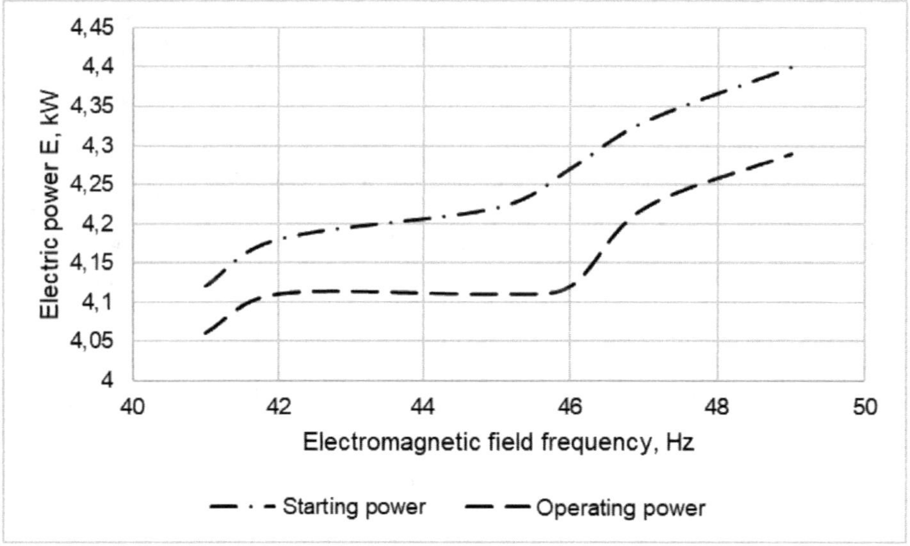

Figure 3. Electric power consumption in LF mode.

As can be seen from Figure 3, the starting power consumption was in the range of 4.12-4.4 kW, and the operating one was in the range of 4.06-4.29. This may indicate that part of the energy in the starting mode is spent on heating the coil and the working chamber of VLA. At the same time, power consumption increases with increasing frequency in a given range.

3.2. Operating frequency mode

For the OF mode, the power consumption was determined in the frequency range 45-80 Hz with a step of 5 Hz. The dependence of the starting and operating power consumption on the frequency of the electromagnetic field is shown in Figure 4.

As can be seen from Figure 4, the starting power consumption was in the range of 2.4-4.48 kW, and the operating one was in the range of 2.37-4.16 kW. This may indicate that part of the energy in the starting mode was spent on heating the coil and the working chamber of VLA, as well as in the LF mode. At the same time, electricity consumption increases with increasing frequency up to 50 Hz. A further increase in frequency leads to a decrease in electricity consumption due to the small amplitude of movement of the ferromagnetic needles. At the same time, starting and operating consumption at 80 Hz is practically equal.

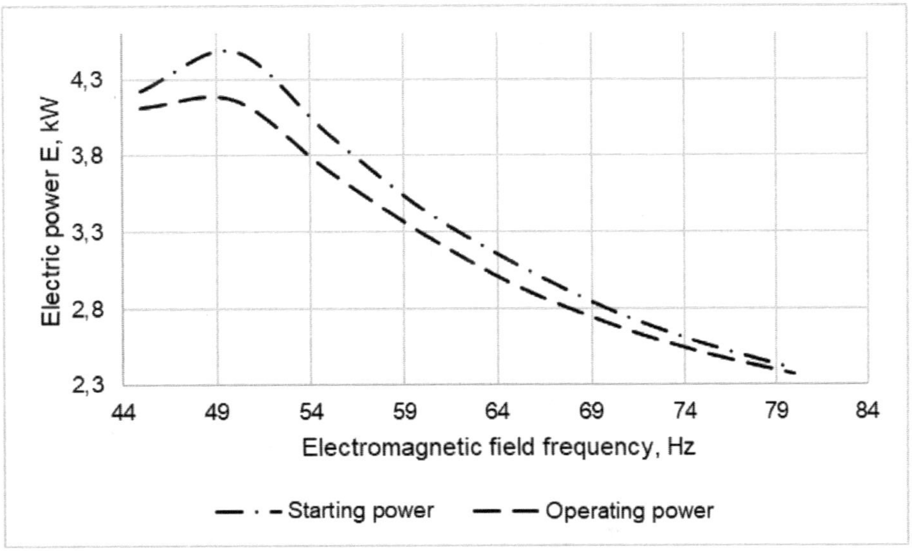

Figure 4. Electric power consumption in OF mode.

Also, in the OF mode, the amount of heat energy removed from the VLA coil by the cooling system was determined to assess the possibility of the subsequent use of cooling water as a source of low-grade thermal energy for the heat pump. Dependencies of heat power in the starting and operating modes on the frequency of the electromagnetic field (OF mode) are shown in Figure 5.

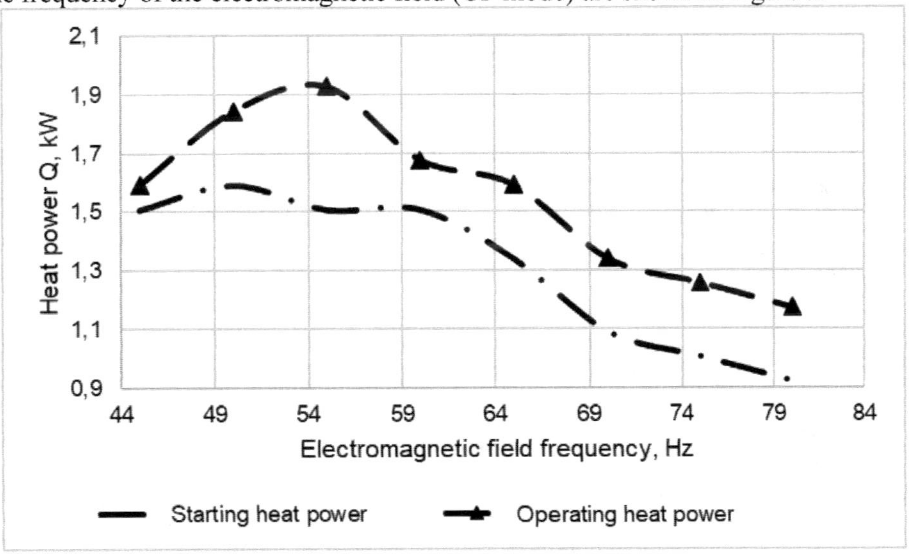

Figure 5. Heat power consumption in OF mode.

As can be seen from Figure 5, the starting heat power was in the range of 0.921-1.591 kW, and the operating heat power was in the range of 1.172-1.872 kW. This may indicate that in the operating mode, thermal energy was not spent on heating the VLA coil and ferromagnetic needles. At the same time, the temperature difference at the inlet and outlet of the water-oil heat exchanger in the starting and operating modes differ by no more than 0.5 °C. The maximum heat power was observed in the operating mode at 55 Hz.

Also, in the course of experimental studies, the coefficient of conversion of electricity into heat on the VLA coil was determined by calculation (equation 2). Figure 6 shows the dependencies of the starting and operating thermal power of the VLA, as well as the coefficient of conversion of electrical energy into heat in the VLA during the starting and operating mode.

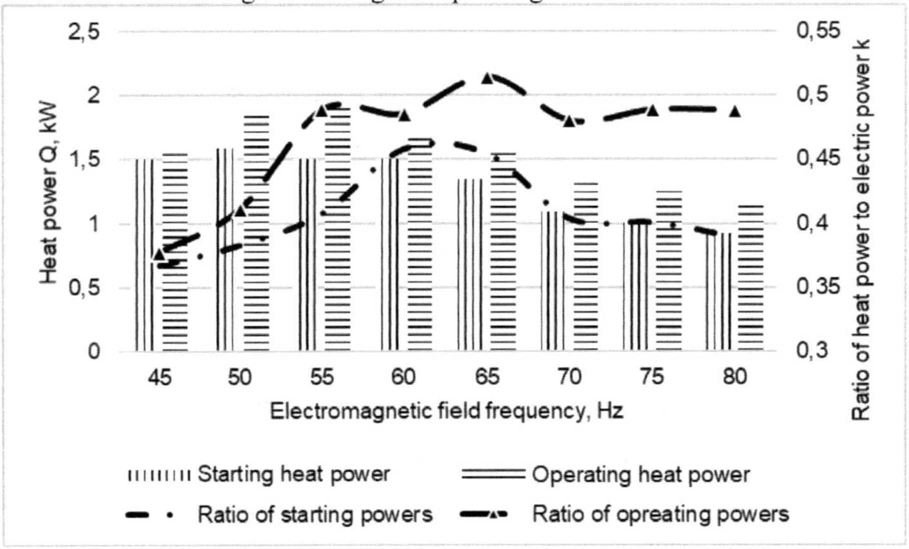

Figure 6. Heat power consumption and the ratio of heat power to electric power in OF mode.

As can be seen from Figure 6, the coefficient of conversion of electrical energy into heat energy is in the range of 0.364-0.458 in the starting mode and 0.377-0.515 in the operating mode. The maximum coefficient of conversion of electrical energy into heat energy was observed at 60 Hz in the starting mode and 65 Hz in the operating mode.

Holik et al. (2021) showed that the overall electrical efficiency of the biogas power plant was increased by 3% when using a heat pump [13]. At the same time, the obtained experimental data on the temperatures of the cooling water make it possible to use VLA as a source of low-grade heat in a number of ways to increase the potential of thermal energy [14-16] with various refrigerants [17], including for the production of electricity [18].

4. Conclusion

Based on the obtained experimental data, it is advisable to use VLA as a source of low-grade heat in the process of waste pretreatment for further anaerobic digestion. At the same time, maintaining the temperature of the cooling water in the range of 10.5-12 °C allows to maintain the operating temperature of the VLA coil while simultaneously obtaining low-grade heat.

In addition, based on the obtained experimental data, the values of the frequencies of the electromagnetic field were determined for further experimental research on the effect of the duration of processing in the VLA:

- 50 Hz, as the mode with the highest power consumption;
- 80 Hz as the mode with the lowest power consumption;
- 55 Hz, as a mode with the highest thermal energy production;
- 65 Hz, as the mode with the highest coefficient of conversion of electrical energy into heat.

Acknowledgments
This work was partially supported by the Federal State Budgetary Institution "Russian Foundation for Basic Research" as part of scientific project No. 18-29-25042.

References

[1] Nozhevnikova A N, Kallistova A Yu, Litti Yu V and Kevbrina M V 2016 *Biotechnology and microbiology of anaerobic processing of organic municipal waste* (Moscow: University book) p 320

[2] Kovalev D, Kovalev A, Litti Yu, Nozhevnikova A and Katrayeva I 2019 Vliyaniye nagruzki po organicheskomu veshchestvu na protsess biokonversii predvaritel'no obrabotannykh substratov anaerobnykh bioreaktorov *Ekologiya i promyshlennost' Rossii* **23(12)** 9–13

[3] Wołosiewicz-Głąb M, Foszcz D and Ogonowski S 2017 Design of the electromagnetic mill and the air stream ratio model *IFAC-PapersOnLine* **50** 14964–14969 DOI: 10.1016/j.ifacol.2017.08.2554

[4] Wołosiewicz-Głąb M, Ogonowski S and Foszcz D 2016 Construction of the electromagnetic mill with the grinding system, classification of crushed minerals and the control system *IFAC-PapersOnLine* **49** 67–71 DOI: 10.1016/j.ifacol.2016.10.098

[5] Khaydarov B, Suvorov D, Poznyak A, Kolesnikov E, Gorchakov V, Mamulat S and Kuznetsov D 2018 Efficient method of producing clinker-free binding materials using electromagnetic vortex milling *Mater. Lett.* **226** DOI: 10.1016/j.matlet.2018.05.016

[6] Khramenkov S V, Pakhomov A N and Streltsov S A 2012 Improving the efficiency of sewage sludge treatment using high-temperature hydrolysis before fermentation *Vodosnabzheniye i sanitarnaya tekhnika* **10** 55–60

[7] Haug R T, Stuckey D C, Gossett J M 1978 Effect of thermal pretreatment on digestibility and dewaterability of organic sludges *J Water Pollut Control Fed* **50(1)** 73–85

[8] Litti Yu, Kovalev D, Kovalev A, Katraeva I, Russkova J and Nozhevnikova A 2018 Increasing the efficiency of organic waste conversion into biogas by mechanical pretreatment in an electromagnetic mill *J. Phys. Conf. Ser.* **1111(1)** 1–8

[9] Kovalev D A, Kovalev A A, Katrayeva I V, Litti Yu V and Nozhevnikova A N 2019 Effekt obezzarazhivaniya substratov anaerobnykh bioreaktorov v apparate vikhrevogo sloya *Khimicheskaya bezopasnost'* **3(1)** 56–64

[10] Kovalev A A, Kovalev D A and Grigoriev V S 2020 Energeticheskaya effektivnost' predvaritel'noy obrabotki sinteticheskogo substrata metantenka v apparate vikhrevogo sloya *Inzhenernyye tekhnologii i sistemy* **30(1)** 92–110

[11] Litti Yu V, Kovalev D A, Kovalev A A, Katrayeva I V, Mikheyeva E R and Nozhevnikova A N 2019 Ispol'zovaniye apparata vikhrevogo sloya dlya povysheniya effektivnosti metanovogo sbrazhivaniya osadkov stochnykh vod *Vodosnabzheniye i sanitarnaya tekhnika* **11** 32–40

[12] Sáez Blázquez C, Borge-Diez D, Martín Nieto I, Farfán Martín A and González-Aguilera D 2021 Multi-parametric evaluation of electrical, biogas and natural gas geothermal source heat pumps *Renew. Energy* **163** 1682-1691 DOI: 10.1016/j.renene.2020.10.080

[13] Holik M, Živić M, Virag Z, Barac A, Vujanović M and Avsec J 2021 Thermo-economic optimization of a Rankine cycle used for waste-heat recovery in biogas cogeneration plants *Energy Convers. Manag.* **232** 113897 DOI: 10.1016/j.enconman.2021.113897

[14] Javad Dehghani M 2021 Enhancing energo-exergo-economic performance of Kalina cycle for low- to high-grade waste heat recovery: Design and optimization through deep learning methods *Appl. Therm. Eng.* **195** 117221 DOI: 10.1016/j.applthermaleng.2021.117221

[15] Wang M, Deng Ch, Wang Y and Feng X 2020 Exergoeconomic performance comparison, selection and integration of industrial heat pumps for low grade waste heat recovery *Energy Convers. Manag.* **207** 112532 DOI: 10.1016/j.enconman.2020.112532

[16] Oluleye G, Smith R and Jobson M 2016 Modelling and screening heat pump options for the exploitation of low grade waste heat in process sites *Appl. Energy* **169** 267–286 DOI: 10.1016/j.apenergy.2016.02.015

[17] Mateu-Royo C, Arpagaus C, Mota-Babiloni A, Navarro-Esbrí J and Bertsch S 2021 Advanced high temperature heat pump configurations using low GWP refrigerants for industrial waste heat recovery: A comprehensive study *Energy Convers. Manag.* **229** 113752 DOI:

10.1016/j.enconman.2020.113752`
[18] He Z, Zhang Y, Wu Z, Ma H and Dong Sh 2017 Experimental study on a bifunctional heat utilization system of heat pump and power generation using low-grade heat source *Appl. Therm. Eng.* **124** 71–82 DOI: 10.1016/j.applthermaleng.2017.05.161

IOP Conf. Series: Earth and Environmental Science **938** (2021) 012005 doi:10.1088/1755-1315/938/1/012005

Feed additive for quails based on fermented brewer's spent grain

N.A. Yurina[1,2], N.D. Labutina[1], A.A. Danilova[1], A.B. Vlasov[1], B.V. Khorin[1], and D.A. Yurin[1]

[1]Krasnodar Research Centre for Animal Husbandry and Veterinary Medicine, Krasnodar, Russia
[2]Kuban State Agrarian University named after I. T. Trubilin, Krasnodar, Russia

E-mail: naden8277@mail.ru

Abstract. The paper discusses the effect of a newly developed complex feed additive Grainbiosorb used in complete feed on livability, age of first egg laying, egg productivity, and feed consumption for the production of 1 dozen eggs of laying quails of the Texas white broiler breed. The scientific experiment was carried out in the experimental vivarium of Krasnodar Research Centre for Animal Husbandry and Veterinary Medicine according to the standard technique recommended by the Federal Scientific Center "All-Russian Research and Technological Institute of Poultry" of Russian Academy of Sciences, (Sergiev Posad, 2013). The developed complex feed additive Grainbiosorb consists of fermented brewer's spent grain, a mineral sorbing complex and probiotic microorganisms. Brewer's spent grain is pre-fermented with probiotic bacterial concentrate, consisting of lactic acid, and propionic acid microorganisms. The use of fermented brewer's spent grain in combination with a mineral complex can solve the problems of metabolism, digestion, exposure to heavy metals and thereby increase productivity and contribute to obtaining safe poultry products. The livability of the quail population in experimental and control groups was the same and amounted to 90.0%. The quails of the first group laid the first egg at the age of 44 days, while the quails of the second group at the age of 42 days. Gross egg production for the period of the experiment per the initial laying quail and the egg production for the average laying quail increased in the experimental group relative to the control by 3.8%. The intensity of egg production in the experimental group increased by 2.3 abs. % relative to that in the control group. Feed consumption for the production of 1 dozen quail eggs decreased by 5.5% in the experimental group, relative to the control.

1. Introduction

To meet the growing market demand for poultry products and the population's demand for high-quality protein, the poultry industry is intensifying through the use of new safe feeds [1]. Probiotics are widely used to improve and maintain human health [2], they are also used in agriculture to improve feed efficiency, stimulate the growth of farm animals, reduce the number of pathogens, and increase immunity [3, 4]. Recently, the use of antibiotics has been minimized and replaced where possible with effective supplements such as probiotics [5].

The animal body is usually detoxified using various enterosorbents, which remove various pollutants that can enter the body with agricultural products [6]. Adsorbents effectively remove salts of heavy metals from the animals' bodies, preventing their intrusion into the final product [7].

During the brewing process, millions of tons of spent grains are produced all over the world, since it is the main by-product of breweries and is formed all year round [8, 9]. Due to its high content of structural carbohydrates, it is a good material for biorefineries. Also, brewer's spent grains contain 20-30% protein and 40-60% coarse fibers [10-12]. It also contains a number of phenolic compounds [13]. In the poultry industry, brewer's spent grains are used as feed to increase feed conversion [14].

The use of modified brewer's grain in combination with a mineral complex can solve the problems of metabolism, digestion, exposure to heavy metals and thereby increase productivity and contribute to obtaining safe poultry products.

This work aims to study the effect of the complex feed additive Grainbiosorb on the livability, egg productivity and feed consumption for a dozen eggs of laying quails of the Texas white broiler breed.

2. Materials and methods

The experiments were carried out in the experimental vivarium of Krasnodar Research Centre for Animal Husbandry and Veterinary Medicine (the village of Znamensky, Krasnodar) using the quails of the Texas white broiler breed according to the standard methodology recommended by the Federal Scientific Center "All-Russian Research and Technological Institute of Poultry" of Russian Academy of Sciences [15].

The method of pairs-analogs was used to divide the laying quails in two groups, 20 heads in each. The quails were kept in brooder cages. The keeping conditions corresponded to zootechnical standards. The experiment scheme is presented in Table 1.

Table 1. The scheme of the experiment

Group	Feed additive
1 - Control	CF (complete feed)
2 - Experimental	98.5 % CF + 1.5 % Grainbiosorb

Complete feed for laying quails in the experiment consisted of wheat, corn, corn germ cake, sunflower cake, soybean cake, meat and bone meal, fish meal, pea concentrate, table salt, soda, limestone flour, amino acids, antioxidants, samples premix of vitamin and enzymes. The compound feed satisfied all the nutritional needs of the quail and its nutritional value did not fundamentally differ between the groups.

Complex feed additive Grainbiosorb consists of modified brewer's spent grain, mineral sorbing complex and probiotic microorganisms. Brewer's spent grain is pre-fermented with probiotic bacterial concentrate, consisting of lactic acid and propionic acid microorganisms.

During the study, the clinical and physiological state of the quail was taken into account by daily examination. The livability of the quail population was calculated as a percentage of mortality from the initial population. The age at which the first egg was laid was recorded. The feed consumption was determined daily, taking into account the production of eggs. The egg productivity of laying quails was taken into account by daily collection and accounting of laid eggs, the weight of eggs was determined using the M-ER 122ACF electronic balance.

The average monthly egg production was processed by the biometric method of variation statistics according to N.P. Plokhinsky [16].

3. Results and discussion

During the first month of laying, 20 eggs were laid per 1 quail in the control group, in the experimental one it was by 5.5% more. During the second month, 20.4 eggs were laid per 1 quail in the control group and 2.0% more in the experimental group. During the third month there were 19.0 eggs in the control group and 3.1% more in the experimental group. During the fourth month 18.1 eggs were in the control group and 5.5% more in the experimental group; during the sixth month 16.1 eggs were in the control group, 3.7% more in the experimental group.

The livability of quails in both groups was 90.0%. Table 2 shows the main economically useful indicators of laying quails.

Table 2. The main economically useful indicators of laying quails

Indicator	Group	
	1	2
Age for laying the first egg, days	44.00	42.00
Gross number of eggs per average laying quail, pcs.	116.40	120.80
Gross number of eggs per group, pcs.	2212.00	2295.00
Average monthly egg production per initial laying quail, pcs.	18.43±0.49	19.13±0.71
Average monthly egg production for an average laying quail, pcs.	19.40±0.38	20.13±0.68
Intensity of oviposition, %	61.44	63.75
Feed consumption for one dozen eggs, kg	0.54	0.51

The quails of the first group laid the first egg at the age of 44 days, while the quails of the second group at the age of 42 days, i.e. 2 days earlier. The gross number of eggs per average laying quail was higher in the second group by 3.8%. The gross number of laid eggs was also higher in the second group. The average monthly egg production for the initial and average laying quail in the experimental group was also higher than in the control by 3.8% but without a significant difference. A positive trend towards an increase in this indicator is observed. The intensity of oviposition was higher in the experimental group by 2.3 abs. %. Feed consumption for the production of 10 quail eggs decreased by 5.5% in the experimental group. The weight of eggs did not differ between groups and amounted to 12.67 ± 0.30 g in the control group and 13.00 ± 0.07 g in the experimental one.

4. Discussion

Throughout the entire experimental period (6 months), there was an increase in the number of eggs in the experimental group relative to the control. Also, the age of laying the first egg in the experimental group decreased by 2 days, which most likely indicates an earlier maturation of the quail. The increase in egg production for quails of the experimental group can be associated with the positive effect of feeding them using the Grainbiosorb complex supplement and the content of additional sources of vitamins, microelements, and biologically active substances in it. Also, feed consumption per ten eggs decreased, most likely due to an increase in feed conversion.

5. Conclusions

As a result of the conducted research, it was revealed that the number of laid eggs increased within 6 months by 2.0-5.5%. The first egg in the experimental group was laid 2 days earlier than in the control. Egg production per initial laying quail (monthly average) and egg production per average laying quail (monthly average) showed a positive upward trend relative to the control by 3.8%. The intensity of egg production is higher in the experimental group by 2.3 abs. % relative to the control group. Feed consumption per ten eggs decreased by 5.5% in the experimental group.

Acknowledgements

The research was carried out within the framework of the state research project FSBSI KRCAHVM No. 0688-2019-0008. We would like to express our gratitude to the staff of the Department of Biotechnology, Biochemistry and Biophysics of the Kuban State Agrarian University named after I. T. Trubilin for assistance in the development of a new complex feed additive Grainbiosorb.

References

[1] Pshatsiyeva Z V, Yurina N A, Khorin B V, Labutina N D, Danilova A A and Ustyuzhaninova T A 2019 Studying the efficiency of natural and synthetic adsorbents in combined feed for broiler chickens *Adv. Anim. Vet. Sci.* **7(s1)** 1–5

[2] Wang H, Lee I-S, Braun C and Enck P 2016 Effect of probiotics on central nervous system functions in animals and humans: a systematic review *Jour. Neurogastroenterol. Motil.* **22**

589
[3] Jeni R E, Dittoe D K, Olson E G, Lourenco J, Corcionivoschi N, Ricke S C and Callaway T R 2021 Probiotics and potential applications for alternative poultry production systems *Poultry Science* **100(7)** 101156
[4] Ramlucken U, Ramchuran S O, Moonsamy G, van Rensburg C J, Thantsh M S and Lalloo R 2021 Production and stability of a multi-strain Bacillus based probiotic product for commercial use in poultry *Biotechnology Reports* **29** e00575
[5] Al-Khalaifah H S 2018 Benefits of probiotics and/or prebiotics for antibiotic-reduced poultry *Poultry Science* **97(11)** 3807–3815
[6] P'yanova L G, Gerunova L K, Likholobov V A, Sedanova A V, Okolelov V I and Lavrenov A V 2016 Development of carbon sorbents based on nanodispersed carbon using biotechnology *Procedia Engineering* **152** 655–663
[7] Tsogoeva F N, Yurina N A, Yurin D A, Kozhokov M K, Baeva Z T and Kalabekov L A 2018 Way to increase digestibility and accessibility of mixed feed nutrients through antioxidants and probiotic supplementation *J. Pharm. Sci. Res* **10(5)** 1192–1194
[8] Budroni M, Mannazzu I, Zara S, Saba S, Pais A and Zara G 2020 Composition and functional profiling of the microbiota in the casts of Eisenia fetida during vermicomposting of brewers' spent grains *Biotechnology Reports* **25** e00439
[9] Parchami M, Ferreira J A and Taherzadeh M J 2021 Starch and protein recovery from brewer's spent grain using hydrothermal pretreatment and their conversion to edible filamentous fungi – A brewery biorefinery concept *Bioresource Technology* **337** 125409
[10] He Y, Kuhn D D, Ogejo J A, O'Keefe S F, Fraguas C F, Wiersema B D, Jin Q, Yu D and Huang H 2019 Wet fractionation process to produce high protein and high fiber products from brewer's spent grain *Food and Bioproducts Processing* **117** 266–274
[11] Sahin A W, Hardiman K, Atzler J J, Vogelsang-O'Dwyer M, Valdeperez D, Münch, Cattaneo G, O'Riordan P and Arendt E K 2021 Rejuvenated Brewer's Spent Grain: The impact of two BSG-derived ingredients on techno-functional and nutritional characteristics of fibre-enriched pasta *Innovative Food Science & Emerging Technologies* **68** 102633
[12] Weiermüller J, Akermanna A, Laudensack W, Chodorski J, Blank L M and Ulbera R 2021 Brewers' spent grain as carbon source for itaconate production with engineered Ustilago maydis *Bioresource Technology* **336** 125262
[13] Martín-Garcíaa B, Tylewicz U, Verardo V, Pasini F, Gómez-Caravaca A M, Caboni M F and Rosa M D 2020 Pulsed electric field (PEF) as pre-treatment to improve the playing quailolic compounds recovery from brewers' spent grains *Innovative Food Science & Emerging Technologies* **64** 102402
[14] PiresFilho I C, Broch J, Eyng C, Silva I M, Souza C, Avila A S, Castilha L D, Cirilo E H, Tesser G L S and Nunes R V 2021 Effects of feeding dried brewers grains to slow-growing broiler chickens *Livestock Science* **250** 104561
[15] Fisinin V I 2013 *Methodology for scientific and industrial research on feeding poultry* (Sergiev Posad: VNITIP) p 33
[16] Plokhinsky N A 1970 *Biometrics* 2nd ed. (Moscow: Publishing house of Moscow State University) p 367

IOP Conf. Series: Earth and Environmental Science **938** (2021) 012006 doi:10.1088/1755-1315/938/1/012006

Comparative analysis of methods for determining the flammability of municipal solid waste as railway transport cargo

L A Koroleva[1], A G Khaidarov[2], G K Ivakhnuk[2], T N Antoshina[1], and A A Boyakhchyan[1]

[1]Saint Petersburg University of State Fire Service of Emercom of Russia, 149, Moskovskiy ave., Saint-Petersburg, 196105, Russia
[2]Saint-Petersburg State Technological Institute (Technical University), 26, Moskovskiy ave., Saint-Petersburg, 190013, Russia

E-mail: lyudamil@mail.ru

Abstract. A statistical analysis of information about fires that occurred on the railroads shows that it is necessary to enhance the fire protection measures while transporting the fire-hazardous materials (FHM). A comparative analysis of the use of the exergetic index and the flammability potential for forecasting the fire risks connected to the railroad FHM transportation has been performed. The studies considered municipal solid waste (MSW) as promising railway cargo. The values of the MSW chemical exergy and the heat of combustion have been calculated. The conditions and behavior of their variations have been investigated. A system of the substances and materials classification that is based on the exergetic coefficient was introduced. The proposed approach allows us to resolve methodological problems caused by the requirement to account for the technical, economical criteria, and the fire risk indexes into a unified classification system. The conception of the exergetic index improves the objectivity of the procedure for the estimation of the MSW flammability properties.

1. Introduction

At present, fire protection measures play an important role. The business activities are based on high technologies, which require an enormous quantity of various resources. This increases the power-consuming, which in turn results in the following challenges: increasing the fire development probability and extending zones of fires, and complicating the techniques for fire extinguishing [1-3].

A statistical analysis of information about fires, and considering the railroad transport features, shows that it is necessary to enhance the fire protection measures while transporting the fire-hazardous materials. One of the perspective techniques is a comprehensive approach based on gathering and processing data that describes the fire risks, and on forecasting their changes. In addition, this approach uses the engineering calculations of characteristics of the fire-protection devices [1,4,5]. On the one side, this approach opens new ways for resolving the fire protection challenges. On the other side, this technique requires high-quality input data and adequate estimations of the fire risk indexes. Simultaneously, those data are input information for engineering calculations [2,6,7] and can be used for the system classification tasks.

A lot of papers [5, 7-10] discuss the methods for determining the fire risk indexes and the flammability characteristics of various substances and materials. However, in most cases, those papers do not consider the following important tasks: improving the efficiency of the used equipment and technologies, and decreasing the environmental pollution. This situation is caused by methodological problems that make it more difficult to account for different characteristics in the comprehensive system [11]. Additionally, an analysis of the fires that occur in the railways shows that the data on the flammability properties of the railroad cargo is not reliable [8].

This work aims to prove the advantages of the exergetic method for estimating the fire risk of the MSW transported by the railways. To achieve this aim, the following tasks have been resolved:

- analyzing the use of the flammability potential as the flammability property of substances;
- establishing relationships between exergy and the fire risk indexes of the railway cargo;
- conducting the numerical and experimental investigations of the exergy, and determining conditions and features of the exergy changes;
- detecting advantages of the exergetic approach for estimation of the fire risks connected to the railroad transportation of MSW.

2. Methods and scenarios of use

The famous Russian scientist V.T. Monakhov made a significant contribution to the scientific background for studying fire safety challenges. To characterize substance flammability, he introduced the flammability potential as a comprehensive fire risk index [5].

The paper by V.T. Monakhov [5] defines the flammability potential ($\Delta\Pi_r$) as an excess or deficit of the flammable system energy. A definition of the flammability potential is based on the thermodynamic investigation of the gaseous mixtures with a minimum capability to inflame. The flammability potential is also connected to values of the mixture components enthalpy (heat function) at the initial temperature and the firing temperature.

The use of the flammability potential $\Delta\Pi_r$ allows us to do the following:

- differentiate the mixtures concerning the $\Delta\Pi_r$ value: flammable ($\Delta\Pi_r < 0$) and nonflammable ($\Delta\Pi_r > 0$) mixture. $\Delta\Pi_r = 0$ characterizes a mixture with a minimum capability to inflame;
- determine a composition of mixtures with the specified flammability value or specified fire-killing capability, or phlegmatizing efficiency;
- calculate the low limit of the inflammability value;
- determine the fire-killing and phlegmatizing concentrations of mixtures.
- However, the use of the flammability potential has some restrictions:
- the potential value depends on the firing temperature. The firing temperature depends on a particular mixture;
- possible chemical transformations of the flegmatizing agent cannot be accounted; the flegmatizing capability is only determined by changes of the agent enthalpy caused by the agent heating;
- the potential value depends on the characteristic dimension of a container where the firing process spreads. This is caused by the effects of the container dimension on the adiabatic firing temperature for a mixture with a minimum capability to inflame;
- it is difficult to calculate the flammability potential for liquids and solid substances because of the specificity of their firing process. In addition, it is not possible to obtain an exact dependence of the flammability potential on the state parameters of substances that are in the condensed state;
- the use of the flammability potential cannot be used to understand the energy-environment efficiency and environmental safety of processes.

The flammability potential is an empirical concept, so it is difficult to determine the exact frames of the applicability of methods based on that conception [9].

The exergetic approach is a promising method for estimations of fire risks [11, 12]. There are a lot of papers that deal with studies of the exergetic efficiency and the environmental compliance of the technical systems and processes [3,13-15]. The implementation of the exergetic approach could be considered as a total solution that is aimed to improve the process energy effectivity and decrease the environmental and fire risks. This approach allows us to overcome methodological problems related to the integration of several parameters into a comprehensive system.

The key concept of this method is exergy. Yugoslavian scientist Z. Rant introduced the exergy to characterize a "capacity for work" [16]. The exergy (e) is similar to Gibbs free energy (G) as in equation (1). On the other hand, the exergy is a more general concept because it depends on the state of both the system and environment, and depends on the ambient temperature T_0 [17]:

$$e = \Delta H - T_0 \Delta S \tag{1}$$

e is the substance exergetic ratio, KJ/mole
ΔH is the difference in enthalpy, KJ/mole
ΔS is the difference in entropy, KJ/(mole*K)
T_0 is the ambient temperature, K.

The exergy depends on parameters of both the system and environment and includes the following parts shown in figure 1.

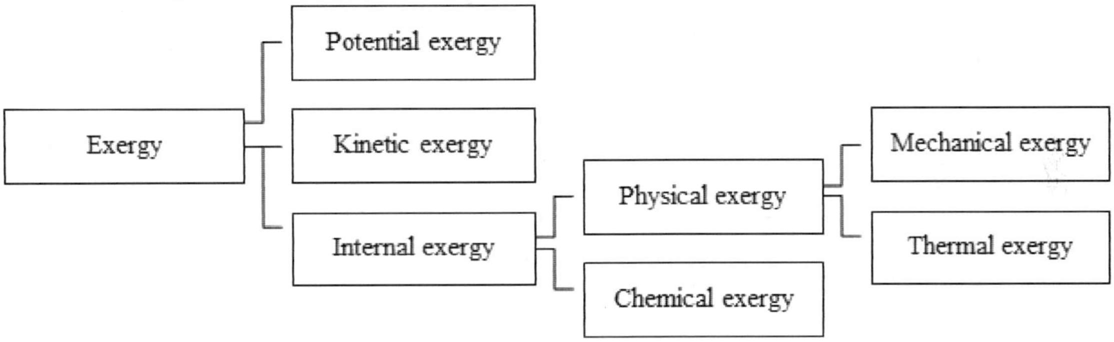

Figure 1. Parts of exergy.

In most real technical systems, the role of the kinetic and potential energies is negligible [16, 17], and they are not considered in this work.

It is possible to consider the exergy as a degree of deviations of the thermodynamic system state parameters from the external environment conditions. Physical exergy describes the differences in pressure (mechanical contribution), and temperature (thermal contribution). Chemical exergy describes changes in the chemical composition at the temperature and pressure that correspond to the external environment parameters.

To prove the practicability of the use of the exergetic approach for estimations and forecast of the fire risks, we consider MSW as promising railway cargo.

Using the information about the MSW morphological composition, we calculated the combustion heat and chemical exergy of the total quantity of MSW and their separate components with the forecast model reported by F. Eboh, P. Ahlström, and T. Richards [12, 18].

The studies are based on a test scenario when 1 kg of the waste is incinerated. Additionally, let us suppose that the waste chemical formula is $C_mH_nN_pO_qC_rS_t$, and the incineration process can be described through the following chemical transformation:

$$C_mH_nN_pO_qCl_rS_t + \left(m + t - \frac{q}{2} + \frac{n-r}{4}\right)O_2 \rightarrow m\,CO_2 + \left(\frac{n-r}{2}\right)H_2O + \frac{p}{2}N_2 + r\,HCl + t\,SO_2,$$

where *m, n, p, q, r,* and *t* are the numbers of the C (carbon), H (hydrogen), N (nitrogen), O (oxygen), Cl (chlorine), and S (Sulphur) atoms respectively.

During the MSW transportation, the environment parameters are thought to be as follows: T_0=298.15 K, and P_0=101.325 kPa.

IOP Conf. Series: Earth and Environmental Science **938** (2021) 012006 doi:10.1088/1755-1315/938/1/012006

3. Results and discussion

The stages of our work were calculations of the chemical exergy, the LHV (Lower Heating Value) of the MSW, and deriving the dependence of those parameters on the waste composition, as shown in figures 2a and 2b.

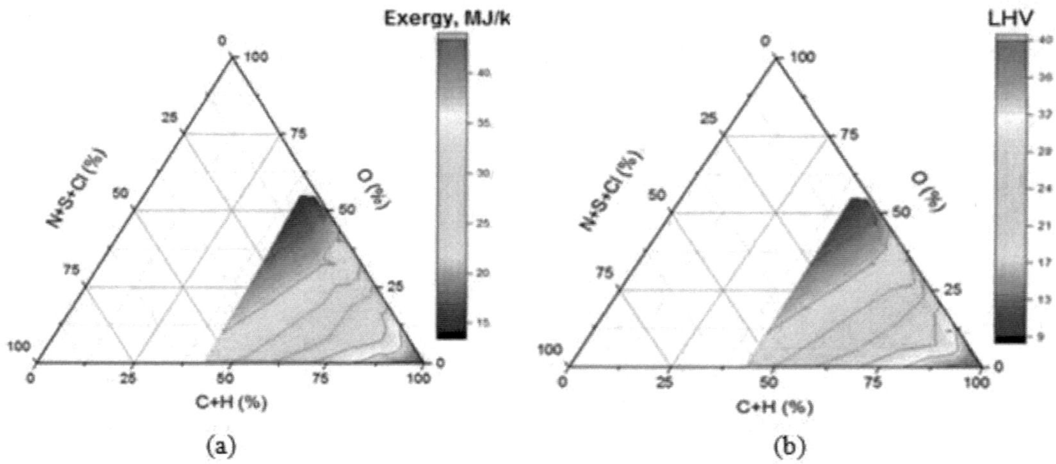

Figure 2. Dependences of the chemical exergy and the low heating value on the MSW composition: a) changes of the chemical exergy; b) changes of the LHV.

We have studied the dependency of the exergy changes on the content of ash A (%), the volatiles V (%), and the fixed carbon FC (%) (see Figure 3). The values of F, V, and FC were obtained using the data from [19]. The following features of the exergy behavior have been revealed: the increase of the volatiles content results in increasing the exergy; the increase of the fixed carbon results in decreasing the exergy; the decrease of the ash-content results in increasing the exergy.

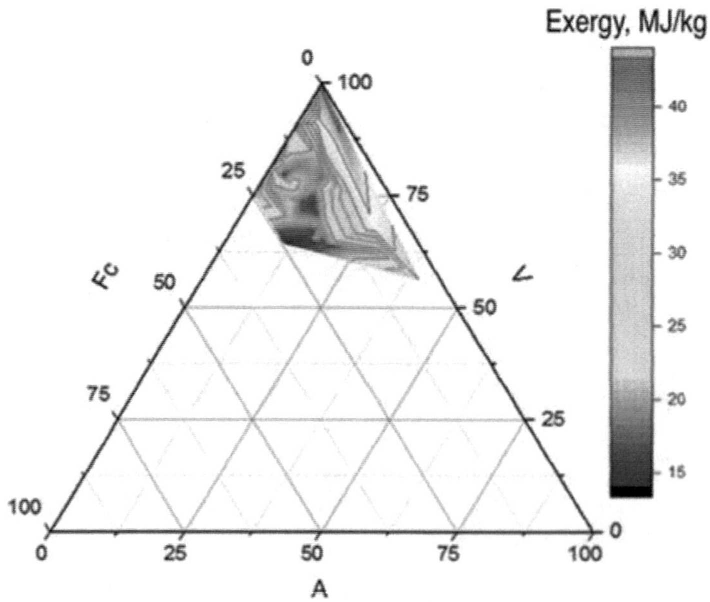

Figure 3. Dependence of the chemical exergy changes on the content of the volatiles V (%), fixed carbon FC (%), and ash A (%).

Table 1 analyses the waste components by LHV and e_{chem} [19].

Table 1. The waste components are sorted by values of the low heating and chemical exergy in descending order.

Parameter	The MSW components listed in descending order
LHV, MJ/kg	Polyethylene> polypropylene> polystyrene> plastic packaging> rubber> acrylic fiber> fish bones> polyethylene terephthalate> polyester fibers> cotton> animal bones> polyvinylchloride > wool> sawdust> walnut shells> wood> wood chips> weeds> fruit peels> toilet paper> cardboard> leaves> printing paper> food starch> vegetables
e_{chem}, MJ/kg	Polyethylene> polypropylene> polystyrene> plastic packaging> rubber> acrylic fiber> fish bones> polyethylene terephthalate> animal bones> polyester fibers> cotton> nut shells> polyvinylchloride> sawdust> wood> wood chips> fruit peels> weeds> toilet paper> cardboard> leaves> food starch> printing paper> vegetables

This analysis shows that the waste can be divided into four groups that include the following components:

1) weeds, fruit peels, leaves, wood, nutshells, vegetables, and the like components that contain pectin, hemicellulose, cellulose, and lignin. The components of this group have a high fixed carbon content and middle values of other indicators;

2) cotton, food starch, potatoes, toilet paper, printing paper, and cardboard that contain repeating units of glucose with different spatial orientation;

3) wool and chemical fibers with low content of H and O, high content of S and N. These components have high values of LHV and e_{chem};

4) plastic and rubber with high values of V (almost 100%), C, H, LHV, and e_{chem}, and low content of O, N, A, and FC.

The obtained pattern lets us prove the use of the exergy for forecasting the fire risks that could be encountered when transporting MSW in railways.

Analysis of the fire characteristics of materials and investigation of relationships between those characteristics and exergy proved that it is useful to consider the so-called exergetic coefficient.

The exergetic coefficient can be used for estimations of the fire risks and environmental hazards of materials and substances as in equation (2):

$$\Pi_e = \frac{e}{30} \qquad (2)$$

e – is the exergetic ratio of the transported MSW (MJ/kg);

30 [MJ/kg] – is the minimum exergy of MSW with the flammability rating E3 (high value of the exergetic coefficient).

The systematization of substances and materials by their exergetic coefficient is shown in figure 4.

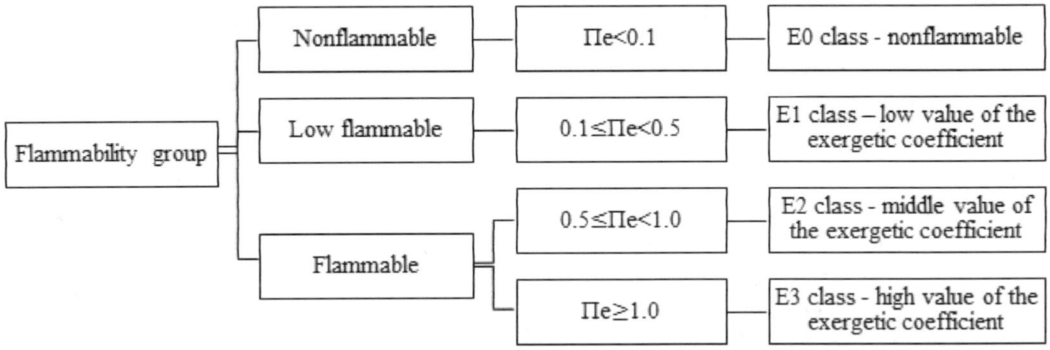

Figure 4. Systematization of the railway cargo materials by the exergetic coefficient value.

To MSW, this systematization is applied in the following way:

- E2 flammability class: wool, chemical fibers, cardboard, paper, food starch, dried leaves, dried grass, dried vegetables, and fruit peels.
- E3 flammability class: plastics and rubber.
- The exergetic coefficient concept improves the procedure for the determination of the MSW fire hazard class and allows us to assign an MSW to a higher class when necessary.

4. Conclusions

A comparative analysis of the use of the flammability potential and the exergetic coefficient showed the following advantages of the latter for estimations of the fire hazard classes of substances, materials, and goods:

1) a possibility to resolve methodological problems caused by mandatory integrating the techno-economical and environmental criteria, and also the fire risk indexes into the comprehensive system;

2) the exergetic coefficient is a quantitative characteristic that depends on both the system parameters and the parameters of the environment, and technological process;

3) a possibility to unify special requirements to hazardous substances contained into the following items: raw materials, goods, semi-products, and waste. These requirements are aimed to ensure fire safety and environmental compliance during the storage, usage, production, and transportation of those items, and estimate a hazard impact of their combustion products;

4) the exergetic coefficient is an additional parameter that helps us select an optimum waste management method. On the one hand, the optimum method improves the objectivity of estimations of fire risks and environmental impacts encountered during waste processing. On the other hand, this strategy allows us to perform a comparative analysis of the exergetic efficiency of methods used for waste processing and recycling.

References

[1] Bariha N, Mishra I M and Srivastava V C 2016 Fire and explosion hazard analysis during surface transport of liquefied petroleum gas (LPG): A case study of LPG truck tanker accident in Kannur, Kerala, India *J Loss Prev Process Ind.* **40** 449–460

[2] Steen-Hansen A, Storesund K and Sesseng C 2021 Learning from fire investigations and research – A Norwegian perspective on moving from a reactive to a proactive fire safety management *Fire Saf J.* **120** 103047

[3] He Z and Weng W 2020 Synergic effects in the assessment of multi-hazard coupling disasters: Fires, explosions, and toxicant leaks *J. Hazard. Mater.* **388** 121813

[4] Vincent C, Corn S, Longuet C, Aprin L, Rambaud G and Ferry L 2019 Experimental and numerical thermo-mechanical analysis of the influence of thermoplastic slabs installation on the assessment of their fire hazard *Fire Saf J.* **108** 102850

[5] Monakhov V T 1979 *Methods for investigating the fire hazards of substances* (Moscow: Chemistry) p 424

[6] Xie W, Zhanga Y, Li J, Mao P and Chen L 2018 Experimental study on characteristics of flame spread over diesel and n-butanol pool fires in tunnel *Tunn. Undergr. Space Technol.* **79** 286–292

[7] Jiao Z, Escobar-Hernandez H U, Parker T and Wang Q 2019 Review of recent developments of quantitative structure-property relationship models on fire and explosion-related properties *Process Saf Environ Prot.* **129** 280–290

[8] Korolchenko A Ya 2015 Problems of determining the combustibility of substances *Fire and Explosion Safety* **24(12)** 6–10

[9] Shebeko Yu N, Ivanov A V and Il'in A B 1982 Concentration of fuel and phlegmatizer at the extremal point of the ignition region *Combust Explos Shock Waves* **18(4)** 411–415

[10] Alexeev S G, Smirnov V V and Barbin N M 2012 Flash point. Part I. Question history, definitions, and test methods of determination *Fire and Explosion Safety* **21(5)** 35–41
[11] Koroleva L A, Khaydarov A G, Ivakhnyuk G K and Koval D N 2018 Exergistic approach to estimation of energy-ecological efficiency and fire hazard of cargo transportation on railway transport *Fire and Explosion Safety* **27(7-8)** 43–52
[12] Khaydarov A G, Koroleva L A and Ivakhnyuk G K 2018 Exergetic assessment of fire hazards of cargo transportation on railway transport *Fire and Explosion Safety* **27(10)** 26–37
[13] Khila Z, Baccar I, Jemel I, Houas A. and Hajjajia N 2016 Energetic, exergetic and environmental life cycle assessment analyses as tools for optimization of hydrogen production by autothermal reforming of bioethanol *Int. J. Hydrog. Energy* **41(39**) 17723–17739
[14] Bellos E, Tzivanidis C and Tsifis G 2017 Energetic, Exergetic, Economic and Environmental (4E) analysis of a solar assisted refrigeration system for various operating scenarios *Energy Convers. Manag.* **148** 1055–1069
[15] Eini S, Shahhossein H, Delgarm N, Lee M and Bahadori A 2016 Multi-objective optimization of a cascade refrigeration system: Exergetic, economic, environmental, and inherent safety analysis *Appl. Therm. Eng.* **107** 804–817
[16] Ghannadzadeh A and Sadeqzadeh M 2017 Exergy aided pinch analysis to enhance energy integration towards environmental sustainability in a chlorine-caustic soda production process *Appl. Therm. Eng.* **125** 1518–1529
[17] Leytes I L, Sama D A and Lior N 2003 The theory and practice of energy saving in the chemical industry: some methods for reducing thermodynamic irreversibility in chemical technology processes) *Energy* **28(1)** 55–97
[18] Eboh F C, Ahlström P and Richards T 2016 Estimating the specific chemical exergy of municipal solid waste. *Energy Sci. Eng.* **4(3)** 217–231
[19] Zhou H, Meng A, Long Y, Li Q and Zhang Y 2014 Classification and comparison of municipal solid waste based on thermochemical characteristics *J Air Waste Manag Assoc.* **64(5)** 597–616

IOP Conf. Series: Earth and Environmental Science **938** (2021) 012007 doi:10.1088/1755-1315/938/1/012007

Engineering-geological and geoecological aspects of formation of dry dumps on hydrodumps

Yu I Kutepov[1], N A Kutepova[1,4], Yu Yi Kutepov[1], A D Vasileva[1], A S Mukhina[1], and R D Smirnov[2]

[1]Saint-Petersburg Mining University, 199106, Saint-Petersburg, 21-line, 2
[2]AO «UK «Kuzbassrazrezugol» «Taldinsky coal mine», Novokuznetsk

E-mail: koutepovy@mail.ru, kutepovyy@ya.ru, nastya-vasiljeva@mail.ru, aleksandra_sergeevna94@mail.ru

Abstract. The paper discusses the issues of increasing the dumping capacity of hydrodumps and their reclamation by filling them with dump embankments from relatively strong overburden. Without the implementation of this measure, the alluvial massifs are inaccessible for a long time for various mining and transport equipment due to the low bearing capacity of the rocks. The specifics of the formation of the composition, state and properties of rocks of alluvial massifs due to the development of fractionation and filtration consolidation processes are noted. The technology of dumping the surface of hydrodump by controlled deformation of the slopes of the first dump layer has been substantiated. A method has been developed for predicting the parameters of intrusion-replacement of alluvial rocks by bulk ones using numerical modeling of the FEM in the Euler formulation. The obtained mechanical properties and regularities of the deformation behavior of technogenic rocks were used to perform geomechanical calculations when justifying the stable parameters of the slopes of the formed natural-technical systems (NTS) "hydrodump+dump" and compaction of embankments to justify the mining-technical stage of reclamation. A geoecological assessment of the impact of the dump NTS on the natural environment has been carried out, which is expressed in the acquisition of land from circulation and the disruption of the natural landscape. The conditions for the formation of technogenic eluvium on the surface of the dump suitable for biological reclamation are considered.

1. Introduction

In open-pit mining of mineral deposits, hydraulic excavation was widely and effectively used at certain stages to remove dispersed rocks composing the upper parts of geological sections. It was explained by the complexity of mining, transportation, and placement in dumps of water-saturated, mostly clay, rocks. At the beginning of the XXI century, the hydraulic excavation application sharply declined due to the high potential hazard of hydraulic structures, the tightening of environmental requirements for the acquisition of land for hydrodumps and the lack of free areas, the low relative capacity of these facilities, poor mobility and the significant cost of the technology used, especially in terms of energy supply, etc.

During the entire period of hydraulic excavation at the open-pit mines in the Kuzbass and mines of the Kursk Magnetic Anomaly (KMA), about 1 and 0.5 million m^3 [1] of dispersed clay and chalk rocks were removed and placed in hydrodumps. The total amount of these hydrodumps was 60 and the total area exceeded 10 thousand hectares, they were located in the immediate vicinity of the allotments. This circumstance predetermines the prospects for the use of alluvial territories for the placement of overburden dumps. These dumps are characterized by increased moisture and porosity, high deformation capacity, low strength, and filtration properties, which persist for a long time (tens of years), therefore the mining and agricultural equipment cannot move along the surface of the alluvial objects. So, required are special studies to address the issues of reclamation of hydrodumps and dumping on their surfaces and the development of original technologies and measures that ensure safety and technical and economic efficiency.

For the first time, hydraulic excavation in Kuzbass was used at the Bachatsky open-pit mine in 1951. Further, other enterprises in the region applied this method. The volumes of hydraulic overburden were constantly increasing, reaching their maximum of 27.25 million m^3 in the mid-80s of the last century. Currently, hydraulic excavation is used only at the Mokhovsky and Kedrovsky open-pit mines. The heights of alluvial objects vary from 4 to 76 m, their occupied area is from 6 to 765 hectares, capacities are from 4 to 200 million m^3. The largest-area alluvial structure in the region is the hydrodump on the Elovka River, the parameters of which (height, area, and capacity) are 60 m, 765 hectares, and 200 million m^3, respectively. The highest is the Bekovsky hydrodump, formed by the reclamation method by the construction of 16 build-up dams up to a height of 76 m. According to this parameter, it is close to the largest hydrodump in Russia (Berezovy Log at KMA), which reached 80 m in height.

At present, 40 aggraded objects are used as the bases of the dumps, which have been dumped since 1970. Many of them have been completely covered by now, but most of them continue to be used for dumping needs.

Analysis of studies on the application of hydrodumps for storing overburden shows that sufficient attention is paid to the issues of geomechanical assessment of the stability of slopes of natural-technical systems (NTS) "hydrodump+dump". However, the geoecological aspects of the formation of such systems have practically not been reflected in the scientific and technical information space. The present research aims to partially fill this gap, as well as to demonstrate new results obtained by the authors during the geotechnical study of the NTS "hydrodump+dump".

2. Materials and Methods

The research is based on an integrated approach, which includes analysis and generalization of the results of previous studies, field and laboratory methods for assessing the composition, state and properties of dump rocks, parameters of exploited technogenic massifs, methods of analytical processing of results, as well as numerical modeling. The issues of dump formation at the territories of the Kuzbass decommissioned hydrodumps, began to be dealt with almost immediately after the end of reclamation of the first structure. The reason for close attention to this facility and the decommissioned dumps is their proximity to open mine workings and the lack of land for the construction of new dumps. In the coal industry, the issues of justifying the use of hydrodump for the placement of dry dumps were discussed by Yu.I. Kutepov [1-6], N.A. Kutepova [2-6], A.V. Mogilin [7], V.P. Zharikov [8], N.G. Fomenko [9], Yu. Yu. Kutepov [6], and others. The works of A.M. Galperin [1,10,11], V.V. Moseikin [1], V.V. Cheskidova [11], S.P. Bakhaeva [12], N. Tamashkovich [13], and others are devoted to engineering-geological assessment of bulk and alluvial technogenic massifs at the mining enterprises of the KMA.

The problem of reclamation of industrial waste disposal facilities is relevant not only for the regions of the Russian Federation. In Europe, urgent is the reclamation of industrial landscapes disturbed by the extraction of coal, peat, and other minerals, including those contaminated with toxic substances. Today Germany is a leader in the development and implementation of reclamation measures [14]. The Loess agreement was concluded in the 60s, in the area of lignite mining (Germany), which required the use of fertile loess for reclamation. For the first time for these purposes, the technology of hydromechanization was applied to reclaim loess on the planned surface of dumps. The study of reclamation of dump massifs continues at present. N. Tamashkovich studies the stability of dams and accumulators of liquid industrial waste in Germany in the long term, subject to the stabilization of areas subject to liquefaction [13], etc.

The need for agricultural reclamation was brought to attention at the end of the 90s in Poland. A biological reclamation experiment was carried out at the coal mine dumps (1997), which used sewage sludge at a dump. Hydroseeding was used to plant the herbal mixture on this layer. It was found that both the sewage sludge and the soil fraction have a positive impact on the characteristics of the reclaimed waste rock [15, 16].

The United States has extensive experience in biological reclamation using hydraulic filling of a seed mixture, where, due to the impossibility of reclaiming disturbed lands, studied was the effect of the mechanical composition of dump soils on the rate of weathering processes. It was shown that erosion processes in sandy soils occur faster than in clayey. The results obtained were used in the storage of pyrite-containing waste with low pH [17, 18].

At present, numerical modeling is used when assessing the stress-strain state of technogenic massifs, including forecasting the processes of replacement-extrusion of alluvial rocks by bulk. The authors have developed an original technique [6] based on the FEM in the Euler formulation, implemented in the Abaqus/Explicit software package. Depending on the initial conditions of the alluvial massif and the accepted load from the weight of the filled embankments, the method makes it possible to obtain changes in the shape and size of the alluvial massif loaded with a dump layer, as well as the relief of the created surface NTS "hydrodump+dump". Data on changes in the macrostructure of the system is the basis for further calculations of the consolidation and stability of the NTS slopes. The results obtained make it possible to substantiate the optimal parameters of structures, including for reclamation.

3. Results and Discussion

3.1. Engineering-geological conditions for the formation of NTS "hydrodump+dump"

When dumps are placed at hydrodumps, new natural and technical systems are formed with very complex engineering, geological and hydrogeological characteristics. Their functioning is not safe in relation to the natural environment and often can pose a threat to mining and infrastructure facilities located in the immediate vicinity. NTS "hydrodump+dump" is a set of interacting soil massifs of different genesis, compositions, and properties: bulk, alluvial, and natural foundations. The first two are technogenic displaced by genesis [13], as they are formed as a result of mining and are associated with the chain of technological processes: "mechanical destruction of overburden - transportation of the formed rock mass to the storage site - placement of its dump or hydrodump". During the formation of such NTS, the state and properties of natural rocks underlying it change as a result of static and dynamic loads.

The processes of formation of technogenic bulk and alluvial rocks are fundamentally different from each when considering their technologies, physical principles, and flow environments. In particular, hydraulic technologies use the destruction of rocks under the action of the kinetic energy of the jet, their transportation by pressure hydrotransport and hydraulic washing into a prepared container. As a result of this technology, alluvial massifs are formed, sometimes of considerable thickness (up to 80 m), mostly composed of poorly permeable rocks of loamy and clayey composition, heterogeneous in plan and section in composition, state, physical and mechanical properties. The established heterogeneity is associated with the development of fractionation processes during pulp filling, i.e., the separation of material by size on the beach, and filtration consolidation, which is expressed in the formation of excessive pore pressure in the rocks. In Kuzbass, when dispersed rocks are injected into hydrodump of on average loamy composition, formed are sandy-sandy loam zones in the alluvial massifs, the length is usually 10-15% of the total length of the structure (L, m), loamy and clayey rocks, having 50- 60% and 30-40% of L, respectively. Zoning is observed in the vertical section of hydrodump, which is associated with the compaction of rocks, while the alluvial massif is divided into subzones of fluid, soft-plastic and refractory rocks, and sometimes (with a high thickness of alluvial rocks) of semi-solid and solid consistency. The subzone of flowing rocks is always confined to the upper part of the section and usually reaches 20-25 m at the moment when the aggrade stops. The bearing capacity of flowing sediments is extremely small and cannot provide the passability of almost any mining and agricultural equipment.

The formation of dry dumps is also determined by several technical and natural factors that we consider within the framework of a single process of technogenesis, which primarily includes the initial lithological composition of the overburden strata. In particular, the dumps in the coal mines of the Kuzbass, working out the seams of the Balakhon series, are dumped mainly with mixtures of sandstone and siltstone, mostly with film-type carbonate cement. To a lesser extent, the structures receive dispersed cover sediments, mudstones and coals. For coal deposits developing coals of the Kolchuginsky series, the advantage of siltstones with clay cement is characteristic. These lithological differences enter the dump in a disturbed state, mix in certain proportions, forming technogenic massifs of rocks, in general, characterized by rather high angles of internal friction (25-320) and significant bearing capacity. This makes it possible to safely move on the dump surface and work with various mining equipment.

However, the filling of dump embankments on a weak alluvial base will almost always be accompanied by the development of dangerous landslide processes. In this case, ensuring safety is possible only through the management and control of deformation processes. Filling of dumps at hydrodump should be considered in two aspects, depending on several restrictive conditions, mainly on the availability of overburden volumes. The first approach is the creation of embankments of a minimum height to ensure the implementation of reclamation works in the absence of overburden material. The second approach is dump filling at stable parameters that ensure the maximum dump capacity of the territory of hydrodump in case of its surplus.

For each of these options, the fundamental issue is the substantiation of the technology for filling the first layer on a "weak" alluvial surface with the maximum and minimum possible height. To choose the height of the dump layer, the technology, and order of dump formation one should study the morphology and mechanisms of geodynamic processes depending on the technological equipment used, namely, bulldozers and excavators (both draglines and shovels) for road and rail transport. In addition, to predict the development of deformations, an appropriate method should be developed depending on the engineering and geological conditions of the hydrodump and the parameters of the filled embankments.

For the case of filling the minimum height of the reclamation embankment, these forecasts are used to determine the required volumes of dump rocks and the subsidence of the embankment surface. Since hydrodumps are characterized by planned heterogeneity in the composition and deformation properties of rocks, it is obvious that the subsidence of the surface of the dump layer is uneven, and in the clay zone, it is much more than in the sandy-sandy loam zone. The change in the dump surface relief should be taken into account when developing measures for the reclamation of this object. In case of substantiation of the maximum parameters of NTS “hydrodump+dump”, its macrostructure obtained at the first stage of filling the dump will further serve as the basis for calculating excess pore pressure in alluvial rocks and stability of slopes depending on the intensity of system formation.

This work considers only the main engineering and geological aspects of the problem of hydrodump utilization for storage of dry dumps. It should be noted that this problem is actually much broader and deeper but the limited number of the cited publications do not allow us to discuss it completely.

3.2. Geoecological problems of the functioning of the NTS "hydrodump+dump"

The main types of impact of dumps on the natural environment are land acquisition for their placement, changes in natural landscapes and, as a consequence, disruption of the functioning of natural biocenoses, engineering-geological and hydrogeological conditions of territories. In some cases, for example, in the presence of harmful components in overburden and stored waste, pollution of nearby territories is observed.

Since during the open-pit mining of mineral resources, large-volume dry dump rocks are formed, the process of land acquisition for their placement is intensified. With the use of hydromechanization, the capacity of hydrodump is much lower and rarely exceeds 0.2 million m^3, therefore hydrodump is the least dumping facility, moreover, very dangerous due to the possibility of a hydrodynamic accident with large environmental damage. Usually, dump structures are located near mine workings in the agriculturally well-developed areas, therefore the very fact of their functioning in the natural environment leaves a certain imprint on it.

A significant negative aspect of the construction of a dump on the allotted land plot is a change in the landscape of the latter, characterized by an initially specific combination of relief, climate, soil, flora, and fauna, etc. The relief can be flat, hilly, and mountainous. Flat relief is characterized by elevation differences up to 25 m and absolute elevations less than+300 m, and hilly relief is characterised by an altitude difference of 25 - 200 m and absolute elevations up to +500 m. The territory with elevations above+500 m belongs to the mountain type. A retrospective analysis of the change in the topography of the territories where the dumps are located on the sections of the Kuzbass shows a drastic evolution from the plain to the mountainous relief. Thus, at the Bachatskiy open pit in the Central Kuzbass in the Sagarlyk gully, hydrodump was reclaimed on an area of 600 hectares with a height difference near dams up to 40 m, which corresponds to the formation of a hilly relief. At present, a dump of dry rocks with a height of more than 100 m is filled on it at an absolute elevation of the surface of+400 m. In the future, it is planned to form a dump of a natural-technical system with a height of 250-300 m with a maximum absolute elevation of a surface of+600 m.

It is known, that relief and climate are interrelated and this relationship can be very diverse. First of all, the climate determines the course of the weathering process, as a result of which a layer of technogenic eluvium is formed on the surface of the dump, and further on the soil, which becomes an environment for the growth of various types of vegetation. The type of vegetation formed on the dump surface, in turn, influences the climate, with different effects noted in the presence of herbaceous and forest vegetation.

For dumps of coal enterprises, the most important factor, both negative and positive, is the presence of coal in the overburden, the content of which in some cases reaches 5% or more. This predetermines the possibility of endogenous fires in the dump massif with the access of oxygen and the development of coal oxidation processes. The temperature in the combustion centers can reach 800-1200 °C, while the mineral and organic parts of the rocks change with the release of hydrocarbons and gases (CO, CO2, NOx, SO_2, and others). The release of carbon monoxide, oxide, and nitrogen dioxide into the atmosphere leads to the adsorption of gases by the soil and a change in the reaction of the medium from slightly alkaline to acidic. Technogenic acidification of soils, in turn, can lead to the sorption of heavy metals. The introduction of coal dust affects the content of soil organic matter, an increase in the number of coal particles, i.e. carbon, which is part of coal dust and soot.

A very important ecological issue in the formation of dumps, taking into account the subsequent reclamation of their surface, is the study of the formation of soils suitable for the growth of agricultural and forestry vegetation. This process depends on the initial granulometric composition of the dump rocks, determining the intensity of its destruction and synthesis of soil-forming substances, their accumulation in technogenic landscapes. The composition of rocks determines the physicomechanical and water properties of the forming soil. The formation of a stable vegetation cover contributes to the retention of soil-forming substances from the removal by water and wind [19]. Dump rocks from the Kolchuginsky and Balakhon deposits are converted into the so-called technogenic eluvium at different rates depending on their strength. The studies of the agrochemical properties of technogenic eluvium on dumps of Kuzbass showed that its properties are quite favorable for the growth of trees and shrub crops.

The formed substrate on the dump surface is characterized by alkaline and strongly alkaline reaction of the soil solution medium (pH_w varies from 6.78 to 8.98). The amount of organic matter in the technogenic soils studied for agrochemical indicators varies from 1.10% to 6.10%. This increased content is associated with the enrichment of the soil mass with carbonaceous particles and coal dust.

The content of mobile phosphorus and nitrate nitrogen in eluvium varies from very low to very high; the gross phosphorus content varies from 0.160 to 0.741%; the total nitrogen content varies from 0.04 to above 0.3%. The presence of coal and moisture in the rocks, as well as an increase in the surface temperature of the dump due to its oxidation processes, the significant carbonate content of rocks, and their increased alkalinity, make some adjustments in the choice of crops, in particular, in favor of plants demanding an increased level of ash nutrition elements in the substrate (birch, linden, maple, etc.) [20, 21].

Analysis of the results of assessing the agrochemical properties of technogenic eluvium on dumps of overburden rocks for their suitability for reclamation (according to the Russian State Standard GOST 17.5.1.03-86) indicates that dump rocks weathered to a sandy loam state correspond to a potentially fertile layer in terms of chemical and granulometric composition, and can be used for biological reclamation of disturbed lands.

4. Conclusion

The considered engineering-geological and geoecological aspects of the formation and functioning of complex natural-technical systems "hydrodump+dump" testify to the unity of these hypostases within the framework of solving the problem of rational use of natural resources and ensuring the safety of dump mining operations. It becomes quite obvious that the reclamation of the territories of hydrodump is practically impossible without applying a layer of overburdened rocks disintegrated in the technological processes of mining on the surface. The thickness of this layer is determined by the presence of sufficient volumes of overburden, allowing, depending on the needs of production, either to form a dump of the maximum possible volume and parameters or to fill an embankment with a minimum height and volume, ensuring the implementation of reclamation work.

In any of the proposed options, the most important technological measure is the initial filling on the alluvial surface of the first layer, which should be carried out in the mode of controlled deformations using the block sequence of the development of the embankment in space. Dump works on each block are carried out until a critical state of deformation behavior appears, after which, they are transferred to the next dump block, etc. To substantiate the value of intrusion-replacement of alluvial rocks by bulk rocks, a method of numerical simulation of the FEM in the Euler formulation is proposed.

The hydrodump massifs are characterized by high heterogeneity in plan and section, caused by the processes of fractionation and filtration consolidation developing during aggrading. These include the formation and dissipation of excess pore pressure in the alluvial rocks with an increase in their thickness. These processes lead to the formation of zones of sandy-sandy loam, loamy and clayey rocks in plan view. In section view, formed are the subzones of different consistency: from fluid to rigid-plastic and even semi-solid and solid formations. The identified zones differ in physical, mechanical, and filtration properties of rocks, as well as their deformation behavior under load. This information is necessary for performing geomechanical calculations when justifying the parameters of embankments and the stability of slopes of the NTS "hydrodump+dump" being formed. It can also be used to predict the subsidence of the dump surface to develop recommendations for the mining and technical reclamation of the object.

According to the geoecological assessment, the main types of adverse impacts of dump NTS on the natural environment include acquisition of significant areas of land from agricultural and forestry industries for the placement of overburden, change of natural landscapes and formation of technogenic landscapes of territories (relief, climate, soil, etc). Compensation of environmental damage caused by dump works involves reclamation with a biological stage. The lack of sufficient volumes of recultivant to carry out this measure predetermines the possibility of using the dumps formed on the surface under the influence of erosion processes of technogenic eluvium. Its investigation on dumps showed agrophysical and chemical properties favorable for the growth of certain types of vegetation.

References

[1] Galperin A M, Kutepov Yu I and et al 2012 *Development of technogenic massifs at mining enterprises* [Osvoyeniye tekhnogennykh massivov na gornykh predpriyatiyakh] (Moscow: Gornaya Kniga Publishing House) p 336

[2] *Recommendations for the engineering-geological substantiation of the parameters of dry rock dumps, filled on the hydrodumps* [Rekomendatsii po inzhenerno-geologicheskomu obosnovaniyu parametrov otvalov sukhikh porod, otsypayemykh na gidrootvalakh] 1985 (Leningrad VNIMI) p 82

[3] *Instructions on the methods of hydrogeomechanical substantiation of the optimal parameters of hydrodumps and dumps on weak foundations. Part II. Substantiation of the optimal parameters of dump structures* [Ukazaniya po metodam gidrogeomekhanicheskogo obosnovaniya optimal'nykh parametrov gidrootvalov i otvalov na slabykh osnovaniyakh. Part II. Obosnovaniye optimal'nykh parametrov otval'nykh sooruzheniy] 1990 (Leningrad VNIMI) p 55

[4] Kutepov Yu I and Kutepova N A 2003 Technogenesis of alluvial deposits [Tekhnogenez namyvnykh otlozheniy] *Geoekologi* **5** 405–413

[5] Kutepov Y I, Kutepova N A, Karasev M A, Vasilieva A D and Kutepov Y Y 2018 Hydrogeomechanical processes in development of spoil dumps and hydraulic fills *Geomechanics and Geodynamics of Rock Masses – Selected Papers from the 2018 European Rock Mechanics Symposium, EUROCK 2018* **(221169)** 1645–1652

[6] Kutepov Yu I, Kutepova N A, Karasev M A and Kutepov Yu Yu 2016 Prediction of deformation of hydraulic-mine dumps overlaid with dump embankment *Gornyi Zhurnal* **12** 23–27

[7] Mogilin A V 2001 *Geotechnical substantiation of the technology for the formation of dump embankments at hydrodumps* [Inzhenerno-geologicheskoye obosnovaniye tekhnologii formirovaniya otval'nykh nasypey na gidrootvalakh] dis....of PhD in Engineering (Moskow)

[8] Zharikov V P 2005 *Engineering-geological and hydrogeological justification of operation and recultivation of hydrodumps of overburden rocks of the Central Kuzbass* [Inzhenerno-geologicheskoye i gidrogeologicheskoye obosnovaniye ekspluatatsii i rekul'tiva-tsii gidrootvalov vskryshnykh porod Tsentral'nogo Kuzbassa] dis....of PhD in Engineering (Moskow) p 186

[9] Fomenko N G 2015 *Geotechnical substantiation of the parameters and technology of dumping at hydrodumps at a high intensity of mining* [Inzhenerno-geologicheskoye obosnovaniye parametrov i tekhnologii otvaloobrazovaniya na gidrootvalakh pri vysokoy intensivnosti gornykh rabot] dis....of PhD in Engineering (Moskow) p 171

[10] Galperin A M, Ferster V and Chef Yu 1997 *Technogenic massifs and environmental protection* [Tekhnogennyye massivy i okhrana okruzhayushchey sredy] MGGU p 534

[11] Moseykin V V, Galperin A M, Cheskidov V V and Punevsky S A 2017 Improvement of remote automated control of slope structures at mining enterprises *Gornyi Zhurnal* **12** 82–86

[12] Bakhaeva S P 2008 *Assessment of the state and forecast of the stability of technogenic soil massifs of coal mines on the basis of integrated monitoring* [Otsenka sostoyaniya i prognoz ustoychivosti tekhnogennykh gruntovykh massivov ugol'nykh razrezov na osnove kompleksnogo monitoring] dis....of Doctor in Engineering (Kemerovo) p 320

[13] Meier T, Wolffersdorff P and Tamaskovics N 2017 Neuartiger Standsicherheitsnachweis für verflüssigungsgefährdete Kippen *Fachsektionstage der DGGTe.V. Würzburg* 1–9

[14] Hawkes G, Gann V, Murphy G and Stephen 2021 Restoration, Reclamation, and Rehabilitation: On the Need For, and Positing a Definition of, Ecological Reclamation *Restoration Ecology* DOI: 10.1111/rec.13461

[15] Kofodziej B, Bryk M, Sfowifska-Jurkiewicz A and et al 2016 Soil physical properties of agriculturally reclaimed area after lignite mine: A case study from central Poland *Soil andTillage Research* **163** 54–63

[16] Pietrzykowski M and Krzaklewski W 2018 Reclamation of Mine Lands in Poland DOI: 10.1016/B978 -0-12-812986-9.0027-0

[17] Swab R, Lorenz N, Byrd S and Dick R 2017 Native vegetation in reclamation: Improving habitat and ecosystem function through using prairie species in mine land reclamation *Ecological Engineering* **108** DOI: 10.1016/j.ecoleng.2017.05.012
[18] Angst G, Mueller C, Angst Š, Pivokonský M and et al 2018 Fast accrual of C and N in soil organic matter fractions following post-mining reclamation across the *USA Journal of Environmental Management* **209** 216–226
[19] Soloviev S, Semina I S, Androkhanov V A and Shipilova A M 2021 Restoration of vegetation cover in reclaimed areas with coal preparation waste in Kuzbass *E3S Web of Conferences* **244** 01015
[20] Bragina P S, Zibert A S and Zavadsky M P 2014 Degradation, restoration and protection of soils. Soils on the overburden dump in the forest-steppe and mountain-taiga zones of Kuznetsk basin. [Degradatsiya, vosstanovleniye i okhrana pochv. Pochvy na otvalakh vskryshnykh porod v lesostepnoy i gorno-tayezhnoy zonakh Kuzbassa] *Soil Science* **7** 878–889
[21] Kopytov A I and Kupriyanov A N 2019 A new strategy for the development of the coal industry of kuzbass and solving environmental problems [Novaya strategiya razvitiya ugolnoy otrasli Kuzbassa I reshenie ekologicheskih problem] *Russian Coal Journal* **11(1124)** 89–93

Modeling of chloroform content in drinking water of infiltration water intake by periods of the annual cycle. Part 1. Extended flood period

M A Malkova[1], A I Vasileva[2], N D Minchenkov[1], and E A Kantor[1]

[1]Ufa State Petroleum Technological University, Kosmonavtov Street, 1, 450062, Ufa,
[2]Center of Analytical Control of Water Quality, State unitary enterprise "Ufavodokanal", Rossyiskaya Street, 157/2, 450098, Ufa,

E-mail physicspoems@yandex.ru

Abstract. A relationship was found between the chloroform content in drinking water and the parameters characterizing the water quality (turbidity, chromaticity, oxidizability) during the extended flood period (March - June), isolated from the eighteen-year period of the infiltration water intake operation. Two time series were formed, including the monthly average values of the parameters for the entire observation period (72 values) and the averaged values of the parameters corresponding to each year (18 values). Two time series were formed, including the monthly average values of the parameters for the entire observation period (72 values) and the averaged values of the parameters corresponding to each year (18 values). Correlation-regression analysis showed that none of the parameters have a decisive significance for the chloroform content. Multiple correlation-regression analysis revealed that the regression equation for the time series of the averaged values of the parameters corresponding to each year is statistically reliable, is characterized by an acceptable mean error (about 12%) and can be used to assess the chloroform content in drinking water.

1. Introduction

Chloroform - a representative of the trihalomethane (THM) group - is a common product resulting from water disinfection. This compound is present in all plumbing systems where chlorine-containing reagents are used to disinfect drinking water [1-5]. The complexity of the ways of chloroform entering the human body predetermines the appearance of increased risks to human health [6-10]. So, in domestic conditions, chloroform affects a person not only enterally, but also through the lungs due to its volatility [4].

Previously, we carried out studies to find a relationship between the THM content in drinking water and the values of generalized and organoleptic indicators: turbidity (M), chromaticity (C) and oxidizability (O) of water source. The choice of indicators is justified by the fact that they indirectly reflect the content in the water source of the main THM precursors - humic and fulvic acids [11]. As a result of the correlation-regression analysis, it was found that the coefficient of determination between the THM and the parameters did not exceed the value of 0.17 [11] when analyzing the entire time period (1997-2014) and slightly increased (up to 0.42) if the analyzed period is determined as the annual average [11]. Thus, the obtained regression equations can be considered as estimates for determining the THM content in drinking water.

Analysis of changes in the annual average monthly values of the selected parameters shows that the maximum THM concentration falls on July, while for the series of oxidizability, chromaticity - in May, and turbidity - in April. Some delay in the maximum of THM content in relation to other parameters of water quality is associated with the following fact. The process of filtering water from a water source is quite long. Taking this fact into account made it possible to model the content of THM using regression analysis for data series shifted relative to each other in such a way that the maximum values of all parameters coincide. In this case, the coefficients of determination were obtained up to 0.80 - 0.91 for an annual period and 0.48 - 0.76 for a time series including 18 years of observation [11].

The following reasons do not allow to obtain an acceptable strength of the relationship between the content of products formed during water chlorination and the parameters that determine the quality of water in a water source:

- firstly, when calculating the strength of the bond between the THM and the selected parameters, a certain error is introduced due to the fact that the THM is a "generalized" parameter. The THM content includes the concentrations of chloroform, bromodichloromethane, dibromochloromethane and bromoform (determined in μg / dm^3), which vary significantly according to the results of drinking water analysis [12, 13]. For example, in 1998 the content of organobromine THMs in relation to the total concentration of THM was 12%, and in 2009 - 56%;
- secondly, previous studies show that the determination of the parameters time series indicates the prevailing value of the random component, which exceeds 80% [11]. In addition, the climatic factor, associated with a shift in the beginning, a change in the duration and intensity of floods in different years, has a great influence.

Chloroform is present in drinking water more often and in higher concentrations than other THMs (bromodichloromethane, dibromochloromethane and bromoform), and is considered as an indicator of the chlorination products content in water [1, 5].

Thus, it may be promising to distinguish from the annual period the segments in which the the water source state depends on various factors. It is known, for example, that the water source quality deteriorates significantly during the flood period [5, 7]. In addition, during the flood spring period, organizations providing water supply to the population are taking a number of technological measures aimed at preserving the quality of drinking water. Such a measure, for example, is an increase in the chlorine dose, which leads to an increase in the chloroform concentration in drinking water [1, 13].

We differentiated the averaged annual cycle and the entire time series into two periods: an extended flood period and a period covering low water and a relatively constant watercourse, and carried out a regression analysis for each of these periods.

2. Materials and Methods

The data for analytical control of the chloroform content in drinking water, obtained by the center for analytical control of water quality of a water supply organization of a large urban agglomeration in a long-term period (1997 - 2014) at an infiltration-type water intake, served as the initial data for the calculations.

For modeling, multivariate correlation-regression analysis was used [14], as a method to reveal the relationship between the content of chloroform (CF) in drinking water with some organoleptic and generalized indicators of water quality of a water source: turbidity (T), chromaticity (C), oxidizability (O).

The measurement of the chloroform content in drinking water is carried out, as a rule, once a month, i.e. 12 measurements are performed per year. Therefore, we calculated the monthly average values of water quality indicators by summing all readings for a month and dividing the resulting value by the number of measurements in a month. Thus, time series of turbidity, chromaticity and oxidizability were formed, containing the same number of measurements as in the series of chloroform concentration values. According to the values of the monthly average values of T, C, O, the average long-term values of these indicators were formed, which makes it possible to obtain 12 values characterizing the averaged year. Such an algorithm for data processing allows maintaining the general trend and dynamics of changes in the parameters T, C, O and CF content within the annual cycle.

In all cases, the quality of the multiple regression equation is assessed according to the following parameters: A - average relative error, %; F - Fisher's criterion; R^2 is the coefficient of determination, S is the standard deviation.

The article presents the accepted designations of parameters related to the characteristics of various periods of the annual cycle: [CF], [T], [C], [O] - the monthly average values of the parameters for the entire observation period (April, May, June, July); $[CF]_{aa}$, $[M]_{aa}$, $[C]_{aa}$, $[O]_{aa}$ are the averaged values of the parameters corresponding to each year (April, May, June, July).

To distinguish an extended flood period, we extracted from the studied parameters the values corresponding to four months (March, April, May, and June), in which the manifestation of flood events and high water is possible. Thus, for processing using correlation-regression analysis, two time series of data on changes in water quality parameters were obtained: the first represents the monthly average values of the parameters for the entire observation period (72 values), the second is obtained as a result of averaging the parameter values corresponding to each year (18 values).

3. Results

When modeling the seasonal variability of water quality, different approaches are used depending on the tasks. Short-term dynamic forecasts are provided, as a rule, by deterministic models, while for long-term tasks aimed at designing or developing a water management strategy, statistical models are needed [14-16]. The main purpose of such models is the statistical assessment of changes in water quality when conditions change in the catchment.

t is known that the chlorine dose is considered as a priority factor providing the formation of THM and, in particular, chloroform [1, 17]. However, modeling the CF content as a parameter that depends on the chlorine dose indicates that it is not always possible to obtain satisfactory results [11]. The reasons for this may be as follows. First, the dose of chlorine (chlorinating agent) is selected to ensure that all microorganisms that can cause serious waterborne diseases are suppressed. Such diseases, for example, include cryptosporidiosis, cyclosporiasis, giardiasis, cholera, dysentery, etc. [1, 18]. Secondly, an additional amount of a disinfecting agent can be introduced into drinking water to ensure a prolonged disinfecting effect when transporting water from the treatment site to the consumer in the case of a significant length of water supply networks. Thus, the amount of the chlorinating agent is three orders of magnitude higher than the amount of formed THMs [11]. In addition, in the system of analytical control of water, as a rule, the determination of the content of humic and sulfonic acids is not provided [16]. On the other hand, the quality of water subjected to chlorination, to a certain extent, determines the qualitative and quantitative composition of drinking water [18, 19].

We have made an attempt to describe the CF content in drinking water obtained at the infiltration water intake using multivariate correlation-regression analysis. The independent variables were water quality indicators of the source: turbidity (M), chromaticity (C), oxidizability (O). The initial data used in the calculations represent the description of the annual cycle, which is the average for the entire observation period (18 years) and the period from 1997 to 2014.

One of the reasons for the high proportion of a random variable in the time series of the CF content in drinking water, as well as for THM [11], may be a shift in the beginning, a change in the duration and intensity of floods in different years. In this regard, it seems appropriate to trace the change in water quality parameters during this period.

The flood period is generally considered to include three months - April, May and June. However, the analysis of water discharge shows that during the flood period it is necessary to include additionally July, since the discharge in this month is increased (Fig. 1).

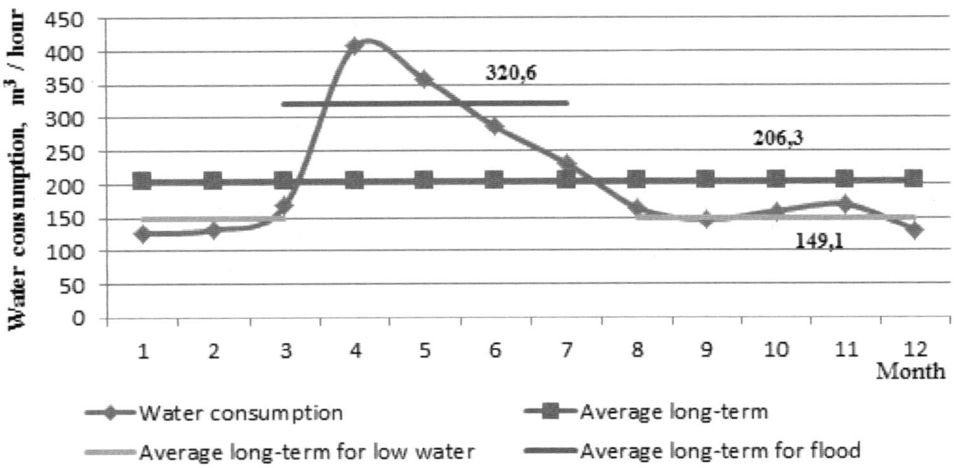

Figure 1. Water consumption in a water source in an annual period (m3 / h)

During the extended flood period from April to July, the water discharge exceeds the average monthly value by 15–97%, and in the period from August to February, the water discharge is lower than the average value by 18–39% (Fig. 2).

Thus, since the state of the water source is largely determined by the flow rate of the incoming water, it is possible to divide the annual cycle into two periods: 1 - extended flood period, 2 - period of low water and relatively constant water flow. In addition, the above data show that the maximum value of the THM content falls on July, while for turbidity - in April, chromatisity and oxidizability - in May. Similar results were obtained when comparing the CF content in drinking water with the parameters characterizing the quality of the water source (Fig. 2).

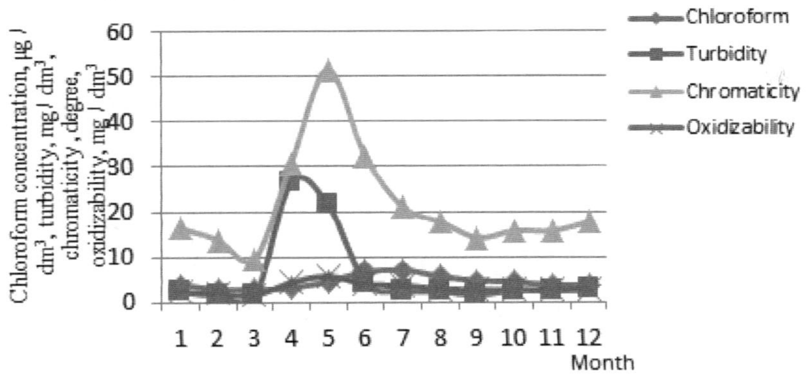

Figure 2. Average long-term values of chloroform concentration, cromatisity, turbidity and oxidizability

In the period from April to July, changes in all analyzed parameters have an inflection point, passing through the maximum values. This circumstance can also serve as a basis for identifying the extended flood (April - July) as specific (with increased stochasticity) in the annual cycle

According to the averaged values of the parameters corresponding to each year, for the extended flood period (April - July), O increases, while CF, T and C decrease (Fig. 3). It should be noted that over an 18-year period (average monthly values of parameters for the entire observation period), the CF content in drinking water varies from 1.5 to 15 μg / dm^3, T - from 1.3 to 63.9 mg / dm^3, C - from 8.5 to 106.9 ^{0}C, O - from 2 to 8.6 mg / dm^3 (Fig. 3).

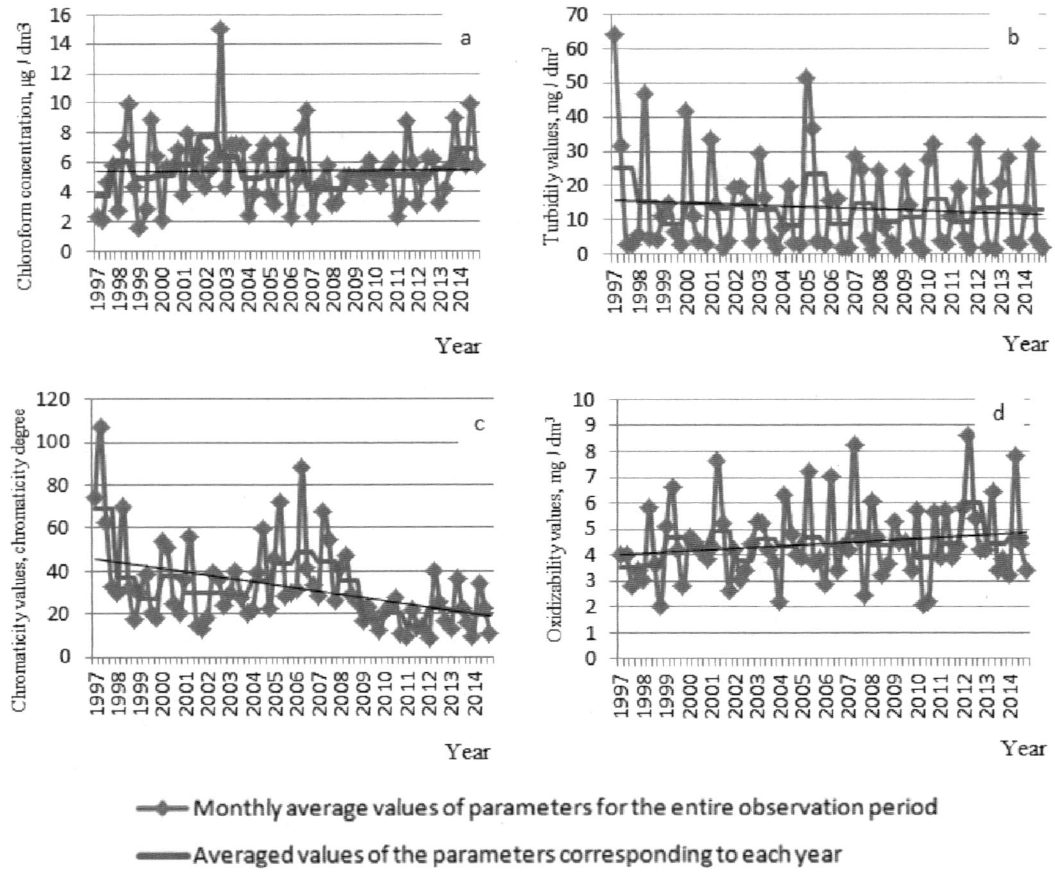

Figure 3. Time series (monthly average and averaged) of indicators in the extended flood period (1997-2014): a) chloroform, μg / dm^3; b) turbidity, mg / dm^3; c) chromaticity, degrees; d) oxidizability, mg / dm^3

Table 1. Equations of dependence of monthly average and averaged concentrations of chloroform [CF] on turbidity [T], chromaticity [C], oxidizability [O]

Monthly average values of parameters for the entire observation period (72 values)					Averaged values of the parameters corresponding to each year (18 values)				
[CF] = 6,10 – 0,02 [T]				(1)	**$[CF]_{aa}$ = 5,83 – 0,04 $[T]_{aa}$**				(4)
A = 52,7	F = 0,21	R^2 = 0,00	S = 4,44		A = 13,4	F = 2,14	R^2 = 0,03	S = 0,94	
[CF] = 7,43 - 0,05 [C]				(2)	**$[CF]_{aa}$ = 6,09 - 0,052 $[C]_{aa}$**				(5)
A = 53,1	F = 2,87	R^2 = 0,04	S = 4,02		A = 13,3	F = 6,73	R^2 = 0,09	S = 0,91	
[CF] = 3,46 + 0,54 [O]				(3)	**$[CF]_{aa}$ = 6,32 - 0,22 $[O]_{aa}$**				(6)
A = 57,0	F = 1,99	R^2 = 0,03	S = 4,37		A = 13,5	F = 1,17	R^2 = 0,02	S = 0,95	

Comparison of the data obtained shows that for a shorter time series, the error is almost four times less, but neither for the monthly average (equations 1-3), nor for the averaged (equations 4-6) values it is possible to obtain a reliable equation, - the values of the obtained determination coefficients vary from 0 to 0.09 (table 1).

The results obtained are primarily determined by the fact that the dominant component for both time series is a random variable, which for the time series of monthly average values is 95.1%, and for the time series of averaged values - 90.8%. Thus, the averaged time series can only be used to estimate the CF content in the extended flood period.

The results of multivariate regression analysis between the chloroform content and parameters for the monthly mean values for the entire observation period (equation 7), for the averaged values corresponding to each year (equation 8) are presented in the table (table 2).

Table 2. Multiple regression equation for the relationship between chloroform [CF] content and parameters: turbidity [T], chromaticity [C], oxidizability [O] during the flood period

Monthly average values of parameters for the entire observation period (72 values)	**[CF] = 4,33 – 0,0004 [T] - 0,07 [C] + 0,87 [O]** A = 58,3 F = 2,59 R^2 = 0,12 S = 4,25	(7)
Averaged values of the parameters corresponding to each year (18 values)	**$[CF]_{aa}$ = 8,40 – 0,007 $[T]_{aa}$ - 0,03 $[C]_{aa}$ - 0,46 $[O]_{aa}$** A = 13,5 F = 3,92 R^2 = 0,17 S = 0,89	(8)

Analysis of equations 7 and 8 indicates that equation 7 is not statistically reliable ($F = 2.59 < F_{cr} = 2.68$), while equation 8 corresponds to ($F = 3.92 > F_{cr} = 3{,}34$); the calculation error by equation 8 is almost four times lower than by equation 7. In both cases, the significance level is $p = 0.05$. Equation 8 is characterized by $R^2 = 0.17$, which corresponds to a correlation coefficient of 0.41, which, according to the Chaddock scale, is defined as a moderate bond strength, which allows us to consider Equation 8 as acceptable for assessing the CF content in drinking water.

It can be assumed that the relatively low strength of the relationship between the CF content and the water quality parameters of the water source is determined by a high proportion of a random variable, which characterizes both the time series of the monthly average values of the parameters for the entire observation period (72 values) and the averaged series of parameter values corresponding to each year (18 values). Determination of the indicated time series leads to practically comparable values of the components - trend, seasonal and random components (table 3).

Table 3. Distributions of the components of the time series of chloroform concentrations,% (symbols: $tr_t + c_t$ - trend-cyclic component; s_t - seasonal component, ε_t - random component

Component	The entire time series	Extended flood period	
		Monthly average values of parameters for the entire observation period (72 values)	Averaged values of the parameters corresponding to each year (18 values)
$tr_t + c_t$	0,9	2,1	2,7
s_t	29,1	22,8	23,0
ε_t	70,0	81,6	74,3

It should be noted that even for a number of averaged values of the parameters corresponding to each year (18 values), the share of the random variable is higher than this parameter, which characterizes the entire observation period (without excluding floods), - 74.3 and 70.0%, respectively. This is one more proof of the high stochasticity of water quality parameters during the flood period.

In general, the results obtained allow us to consider the found equations as estimates, which can be considered as statistical models suitable for solving long-term problems and of interest in the design or development of a water management strategy.

4. Conclusion

The analysis of predicting the chloroform content according to the general and organoleptic indicators of the water source quality: turbidity, chromaticity and oxidizability using correlation-regression analysis during the extended flood period (March-June) was carried out. It was found that the chloroform content cannot be determined by the quality parameters of the water source using pairwise ratios, which are determined by low values of the determination coefficients. We have formed two time series: from the monthly average values of the parameters for the entire observation period (72 values); and the average values of the parameters corresponding to each year (18 values). Using the equations of multiple correlations between the chloroform content in drinking water and the water quality parameters of the water source, a moderate bond strength was determined (in accordance with the Chaddock scale), which allows us to consider the equations found as estimates. The equations can be considered statistically reliable and suitable for solving long-term problems and of interest in the design or development of a water management strategy.

Acknowledgments
The work was performed within the framework of the Ministry of Science and Higher Education of the Russian Federation state task in the field of scientific activity, publication number FEUR - 2020 - 0004 «Solving urgent problems and researching processes in petrochemical industries accompanied by flows of multiphase media»

References

[1] Goncharuk V V 2014 *Drinking water: physics, chemistry and biology* (Springer International Publishing)
[2] Rook J J 1974 Formation of haloforms during chlorination of natural waters *Water Treatment Examination* **23** 234–243
[3] Ademollo N, Patrolecco L, Polesello S, Valsecchi S, Wollgast J, Mariani G and Hanke G 2012 The analytical problem of measuring total concentrations of organic pollutants in whole water *Trends Anal. Chem.* **36** 71–81
[4] Shan J, Hu J, Kaplan-Bekaroglu S, Song H and Karanfil T 2012 The effects of pH, bromide and nitrite on halonitromethane and trihalomethane formation from amino acids and amino sugars *Chemosphere* **86** 323–328
[5] Lantagne D S, Cardinali F and Blount B C 2010 Disinfection by-product formation and mitigation strategies in point-of-use chlorination with sodium dichloroisocyanurate in Tanzania *Am J Trop Med Hyg* **83(1)** 135–143
[6] Maikova M A, Vozhdaeva M Y and Kantor E A 2018 Assessment of carcinogenic risk to population health due to the quality of drinking water of surface and infiltration water intakes *Water and Ecology* **1** 59–64
[7] Malkova M A, Vozhdaeva M Y, Kantor E A and Melnitskii I A 2019 Comparison of drinking water quality in terms of relative concentrations and carcinogenic risks of trihalomethanes *IOP Conference Series: Earth and Environmental Science* **315(5)** 052052
[8] Bulla R, Reckhow D, Li X, Humpaged A, Joll C and Hrudeyc S 2001 Potential carcinogenic hazards of non-regulated disinfection by-products: Haloquinones, halo-cyclopentene and cyclohexene derivatives, N-halamines, halonitriles, and heterocyclic amines *Toxicology* **286** 1–19
[9] Calderon R L 2000 The epidemiology of chemical contaminants of drinking water *Food Chemical Toxicology* **38** 13–20
[10] Nieuwenhuijsen M J, Frellier J, Smith R, Iszatt N, Bennett J and Best N 2009 The epidemiology and possible mechanisms of disinfection by-products in drinking water *Philos Transact Math Phys Eng Sci.* **367(1904)** 4043–4076
[11] Malkova M A, Vozhdaeva M Y and Kantor E A 2020 Method for predicting trihalomethanes content in drinking water (by the example of surface type water intake) *IOP Conference Series: Materials Science and Engineering* **775(1)** 012092
[12] Vozhdaeva M Y, Cypysheva L G and Kantor L I 2005 Efficiency of a combination of mass-selective and atomic-emission detection in the chromatographic analysis of water quality *Mass-Spektrometriya* **2(3)** 229–235

[13] Vozhdaeva M Y, Cypysheva L G and Kantor L I 2001 Analiz organicheskih zagryaznitelej vody metodami gazovoj hromatografii s razlichnymi vidami detektirovaniya *Analitika I kontrol'* **5(2)** 171–185
[14] Larry Wasserman 2014 *All of Statistics* (Springer, New York, NY)
[15] Dolgonosov B M, Vlasov D Y, Dyatlov D V, Suracheva N O, Grigoryeva S V and Korchagin K A 2004 characteristics of the water quality variability in case of water supplied to a pumping station *Inzhenernaya ekologiya* **3** 2–20
[16] *Drinking Water. Hygienic Requirements for Water Quality of Centralized Drinking Water Supply Systems* SanPiN 2.1.4.1074–01. Pit'evaya voda
[17] Miftahova K R, P'yankova O G, Rudakova L V and Glushankova I S 2015 Hlorirovanie kak osnovnoj metod obezzarazhivaniya pit'evoj vody *Ekologiya i nauchno-tekhnicheskij progress. Urbanistika* **1** 233–242
[18] Vozhdaeva M Yu, Holova A R, Vagner E V, Truhanova N V, Mel'nickij I A, Mullodzhanov T T and Kantor E A 2021 Changes in the indicators of the chemical safety of drinking water in Ufa during its transportation to consumers *Gigiena i sanitariya* **100(4)** 396–405
[19] Vozhdaeva M Yu, Holova A R, Mel'nickij I A, Belolipcev I I, Vozhdaeva Yu S, Kantor E A and Lebedev A T 2021 Monitoring and statistical analysis of formation of organochlorine and organobromine compounds in drinking water of different water intakes *Molecules* **26** 18

Metal chips preparation for utilization using advanced reagents

A Samukov[1], M Cherkasova[1], M Kuksov[1], and S Dmitriev[1]

[1]REC "Mekhanobr-Tekhnika", St. Petersburg, 22nd line, 3, bldg. 5, Russia

E-mail: samykov_ad@mtspb.com, cherkasova_mv@mtspb.com, kuksov_mp@mtspb.com, dmitriev_sv@mtspb.com

Abstract. The article touches upon the problems of utilization of metal chips containing residues of cutting fluids. The research results are shown for nickel alloy chips washing to remove cutting fluids with the use of various reagents available on the current market. In the course of the research, an original colorimetric express method for assessing the contamination of metal chips was developed, which allows quickly and efficiently establishing the quantitative drop in the cutting fluids content in the chips after the washing. Conclusions are made as to the most effective reagent from the point of view of economics and quality of cutting fluid removal.

1. Introduction

Layer-by-layer laser-based metal powder fusion is the most advanced developing and innovative technology for the manufacture of parts from metals or alloys [1, 2]. The major advantage of this 3D printing technology consists in the ability to manufacture parts with minimum waste generation. Note that, in traditional manufacturing by mechanical processing of blanks, up to 40% of the metal reports to waste in the form of chips, mostly represented by valuable alloys containing expensive alloying components. Their extraction and production are quite energy-intensive, which requires their efficient consumption [3].

Nevertheless, the traditional metalworking method is still used globally. The Russian metalworking industry alone produces millions of tons of metal chips annually, out of which around a quarter is represented by alloy steel and non-ferrous metals. These are mainly utilized through remelting, during which up to 20% of alloying additives burn out and are lost irretrievably. Metal chip utilization without remelting, with the use of mechanical grinding to obtain metal powders, may have significant economic and environmental benefits due to the recovery of such valuable alloying additives, reduced power consumption and production of the most valuable and expensive raw materials for the more advanced technologies in mechanical engineering [4, 5].

However, direct grinding of metalworking waste and subsequent use of metal powders in additive manufacturing are hindered by the presence of cutting fluids in the chips. Cutting fluids are added to remove heat from the metalworking tool by forming a film on the surface, as well as to prevent the resulting chips from sticking to the tool. Cutting fluids may vary widely in composition and most often include the following components: oleic acid; emulsion solutions of sodium nitrite and soda ash with solid lubricants (paraffin, wax); mineral, animal, and vegetable oils with additives of phosphorus, sulfur, and chlorine (sulfofresols); kerosene; and surfactants [6, 7].

The amount of residual cutting fluids covering the chips discharged from metalworking machines may reach 30% of the weight of the chips. When chips with an admixture of cutting fluids are briquetted or bundled in a cold state, the density of the resulting bundles and briquettes is 10–15% lower than that of processed dry chips, free of all cutting fluids. The presence of cutting fluids in the metalworking waste during grinding leads to the formation of refractory ash deposits in the grinding product, which increases the content of sulfur and non-metallic impurities, while the decomposition of hydrocarbons saturates the metal with hydrogen, thereby embrittling it, and the emissions of organic degradation products cause environmental pollution [8, 9].

The following methods are traditionally used to remove impurities and cutting fluids from metalworking wastes: heating in furnaces of various configurations; or centrifugation or washing with special reagents followed by drying [10–12]. Washing is the simplest of these methods since it does not require any specially manufactured equipment and allows using almost any industrial mixers (agitators) that are widely available on the market. This method is extensively used for degreasing large metal parts for subsequent application of various coatings. It is also suitable for washing pre-crushed metal chips, while the applications of the centrifugation method are limited by the high probability of clogging of the perforations in the centrifuge working surface and the ingress of fines into the cutting fluid discharge. The heating method is poorly suited for metal chip cleaning due to its high energy consumption and to the settling of resins and other combustion products on the particle surface, which has an extremely negative effect on the subsequent grinding process [13–15].

The range of industrial detergents available on the current market is quite extensive, which allows searching and selecting the best reagent for the removal of cutting fluid residues from metal chips.

2. Materials and Methods

The most advanced, common, and widely advertised industrial detergents, such as Bast-Klin and Reiniger (HD Reinige), the degreasers for surfaces and equipment recommended by GOST 9.402-2004 (ISO 27831-1:2008), such as TMS Univeco and Vertolin, and manually prepared aqueous-alkaline solutions (with reagent-grade NaOH and distilled water) were selected for the research. The manufacturers of the products do not represent the exact formulations, but all products belong to alkaline degreasers. The “reference” result with the best cleaning performance was obtained when using the nefras organic solvent.

The crushed chip washing tests using nickel alloy chips (77% Ni) were conducted using a tightly sealed plastic cylinder with the diameter of 84 mm and length of 110 mm. A chip sample weighing 100 g and 150 ml of the test detergent were added into the cylinder, jointly taking up about 30% of its volume. The cylinder was installed onto driven rotating rollers, inducing rotation at 76% of the critical rate to ensure the tumbling mixing regime. After 30 minutes of washing, the liquid phase was drained and 150 ml of distilled water was added to the cylinder, after which the cylinder was rotated on the rollers for another 15 minutes. The chip sample was then discharged from the cylinder onto a cotton filter cloth, while being additionally manually washed with approximately 850 ml of the distillate. After the discharge and washing, the chips were left to dry in natural conditions for 12 hours.

The cutting fluid removal quality was assessed using various methods: by weight, visually, by a luminescent assay, and by wetting with water. A visual inspection is the most efficient and cost-effective method for monitoring the residual contamination. It implies wiping of the metal surface with a white cloth or filter paper and then comparing the contamination with a conventional five- to ten-point scale.

Not being of a quantitative nature, however, this method fails to ensure the required accuracy, since visual perception is subjective and imperfect. The highly expensive gas chromatography method is currently used to quantify the cutting fluids content. In this case, however, it is suggested to upgrade the visual method using simple office equipment. Oil prints on paper may be easily scanned and analyzed in a graphics editor, obtaining quantitative chromaticity values characterizing the content of cutting fluids before and after the washing, after which the removal efficiency may be calculated.

IOP Conf. Series: Earth and Environmental Science **938** (2021) 012009 doi:10.1088/1755-1315/938/1/012009

Based on this idea, additional studies were carried out to develop a quantitative express analysis method for establishing the cutting fluids content in metal chips.

Since a flatbed scanner is most efficient for flat research objects, a CanoScan LiDE 70 scanner was used to obtain digital images of cutting fluid prints on paper sheets.

Size A4 80 g/m2 office paper was used for the oil prints. Experiments with thicker paper (220 g/m2) render unsatisfactory results due to the insufficient contrast of the prints.

In order to simplify the digital processing and analysis of the images, the monochrome (black-and-white) scanning mode was selected, with only the scanning threshold parameter set for the tests. This parameter controls conversion of the original grayscale (256 shades) image to monochrome (2 shades) by comparing the brightness of each point with the threshold value, above and below which the points are considered white or black respectively. The black content in the test areas of the samples scanned was determined using Adobe Photoshop and GIMP graphics editors.

It is critical to establish the optimal scanning threshold to obtain the best contrast for different prints. When the scanning threshold is minimized, the "background" black on blank paper will tend toward zero, while its content in the oil print will contrast very slightly with other prints. And vice versa, high scanning thresholds will significantly increase the "noise" level on the scanned image, which will again cancel the differences between the oil print scans.

In order to establish the correct threshold, test scanning of several different research objects was carried out in the range of threshold values between 195 and 218 (see Table 1). Scans were studied for prints of the original material (unwashed chips) and of the material washed using various detergents. Based on the resulting maximum contrast values (the difference between the samples before and after the washing minus the background values), the weighted average scanning threshold was calculated at 210.42, with the nearest integer values of 210 and 211.

Table 1. Average contrasts between the samples after washing with detergents and the original sample minus the background values, depending on the scanning threshold

Scanning threshold	Contrasts after washing with detergents, % (black content)			
	Vertolin	Reiniger	Bast-Klin	TMS Univeco
195	9.8	9.5	9.8	9.8
196	11.1	10.8	11.1	11.1
197	12.3	11.9	12.3	12.1
198	14.4	13.8	14.3	14.0
199	16.0	15.3	15.8	15.4
200	17.9	16.9	17.5	16.8
201	20.0	18.9	19.5	18.6
202	22.4	20.8	21.5	20.2
203	24.3	22.4	23.1	21.1
204	26.9	24.2	25.1	22.4
205	29.3	25.9	26.8	23.3
206	32.3	28.0	28.9	24.4
207	34.8	29.6	30.3	24.8
208	36.4	30.2	30.5	24.1
209	38.2	31.1	30.9	23.3
210	39.9	31.5	30.9	22.4
211	41.5	31.7	30.7	21.6
212	42.0	31.1	29.4	19.7
213	42.8	31.1	28.5	18.4
214	42.7	30.4	27.0	17.2
215	42.5	29.8	25.1	16.6
216	41.1	28.3	22.9	14.5
217	39.7	27.1	21.0	13.7
218	37.8	25.2	19.0	12.9

The analysis of the results obtained shows that the flatbed scanner has a certain scatter of background values for each given threshold in terms of the black content, and this scatter is more intensive with higher scanning threshold values (see Figure 1). Consequently, it would be inadequate to focus solely on the use of the optimal scanning threshold to ensure representative results. In order to take into account the above deviations, the samples were additionally scanned while reducing or increasing the scanning threshold in the ranges above and below a certain background threshold value. The desired background black content threshold was obtained by calculating the weighted average value for the results at the scanning thresholds of 210 and 211. The resulting calculated threshold value for the background is 6.48%, rounded up to the nearest integer for convenience (to 7%).

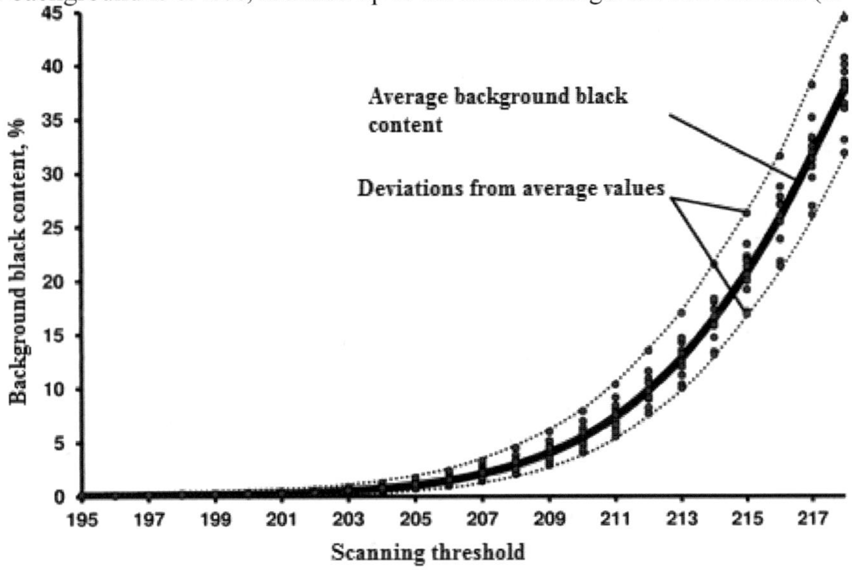

Figure 1. Average black content in the background and deviations from average values depending on the scanning threshold

The actual application of this method to obtain scans with a background close to the threshold value has indicated that an average of two, but not more than three scans must be made for each print, which is an acceptable complication of the method in general.

In addition to determining the optimal values for the scanning threshold and background, the following were also studied during the research:

- The progressive phenomenon of loss of black content in an oil print, associated with partial evaporation of cutting fluid traces from the surface of the paper. The manifestation of this effect limits the time frame for using the oil prints obtained, since their scanning on the second day (or later) will increase the measurement error;
- The effect of desynchronization of the results when using two scanners of the same brand and series, indicating that each scanning device has its own unique settings for the scanning elements, which cannot be synchronized. Consequently, non-parallel application of several scanning devices without the subsequent correction of the deviations shall be unacceptable.

To summarize the positive experience gained in the research, the main provisions for the colorimetric express analysis method are formulated below.

- Metal chip samples weighing 100 g are used to analyze the quality of cutting fluids removal. In order to increase the measurement accuracy, the samples are divided into four parts, evenly laid out using a stencil (70 × 70 mm) in the centers of paper sheets with the density of 80 g/m2 and left for two days to obtain respective oil prints.

IOP Conf. Series: Earth and Environmental Science **938** (2021) 012009 doi:10.1088/1755-1315/938/1/012009

- Within 24 hours after the receipt, the oil prints are scanned using a flatbed scanner to obtain monochrome (black and white) images with the resolution of 600 dpi. The black content for each oil print (within an area of 1520 × 1520 px (pixels)) is determined using Adobe Photoshop or GIMP graphics editors. In order to measure the "background" values for the sheet scan, similar areas (1520 × 1520 px) are selected above and below the oil print, where the black content is calculated with subsequent averaging of the results. The scanning begins with setting the sensitivity threshold of 210 in the scanner software. The average background black content must be as close as possible to 7%, but must not exceed it. If the result is higher or significantly lower, the sensitivity threshold must be reduced or increased, respectively, and a repeated scanning must be conducted. The parallel values obtained for the prints of a single sample must be averaged.
- In addition to the test samples, a print of the original (unwashed) chip sample must be scanned. If necessary, a reference sample (with the maximum effect) may be scanned, washed using, for example, the Nefras organic solvent.
- Based on the averaged black content values obtained for the prints, the reduction in the cutting fluids content is calculated as a percentage of the initial level using the following formula:

$$\varepsilon = 100 - 100(\beta \div \alpha),$$

where: ε is the reduction of the cutting fluids content in the sample, in % of the initial level; and β, α are the black content values for the prints of the washed and original samples, respectively, minus the noise, %.

The analysis of the values obtained is deemed sufficient to assess the quality of cutting fluid removal from the chips.

3. Results and Discussion

Using the calculated values of ε, it is then possible to approximately calculate the mass fraction of the cutting fluids in the washed chips. For this purpose, their contents in the initial and "reference" samples were preliminarily established using gas chromatography at 1.6 and 0.13 g/kg, respectively. These data were used to calculate the corrected value of ε (εEcorr) for the reference sample of metal chips washed with the Nefras organic solvent according to the formula:

$$\varepsilon Ecorr = 100 - 100\ (\beta E \div \alpha isx),$$

where: εEcorr is the corrected reduction of the cutting fluids content in the reference sample, in % of the initial level;

βE is the mass fraction of cutting fluids in the reference sample of washed chips, g/kg; and

αref is the mass fraction of cutting fluids in the initial chip sample, g/kg.

$$\varepsilon Ecorr = 100 - 100 \times (0.13 \div 1.6) = 91.9\%$$

The εEcorr value (91.9%) calculated on the basis of the gas chromatography data differs from the εE (94.9%) calculated from the express colorimetric analysis data. Therefore, it is necessary to proportionally correct the values of ε (εcorr) for the remaining values (see Figure 2) according to the formula:

$$\varepsilon corr = (\varepsilon \times \varepsilon Ecorr\ )/\varepsilon E,$$

where: εcorr is the corrected reduction of the cutting fluids content in the sample, in % of the initial level; and εE is the reduction of the cutting fluids content in the reference sample (uncorrected), in % of the initial level.

The corrected values of εcorr are used to establish the mass fraction of the cutting fluid in the washed samples (see Figure 2) according to the formula:

$$B = \alpha ini \times (100 - \varepsilon corr)/\ 100,$$

where: B is the mass fraction of cutting fluids in the washed samples, g/kg.

The use of the colorimetric express analysis method to monitor the contamination of metal chips with cutting fluids allows assessing detergent effectiveness, however a truly optimized reagent selection requires a comparative study between the cleaning effectiveness and economic feasibility. A diagram was designed (see Figure 3) to illustrate the reduction in the cutting fluids content in the sample when using a specific detergent in comparison with its specific cost.

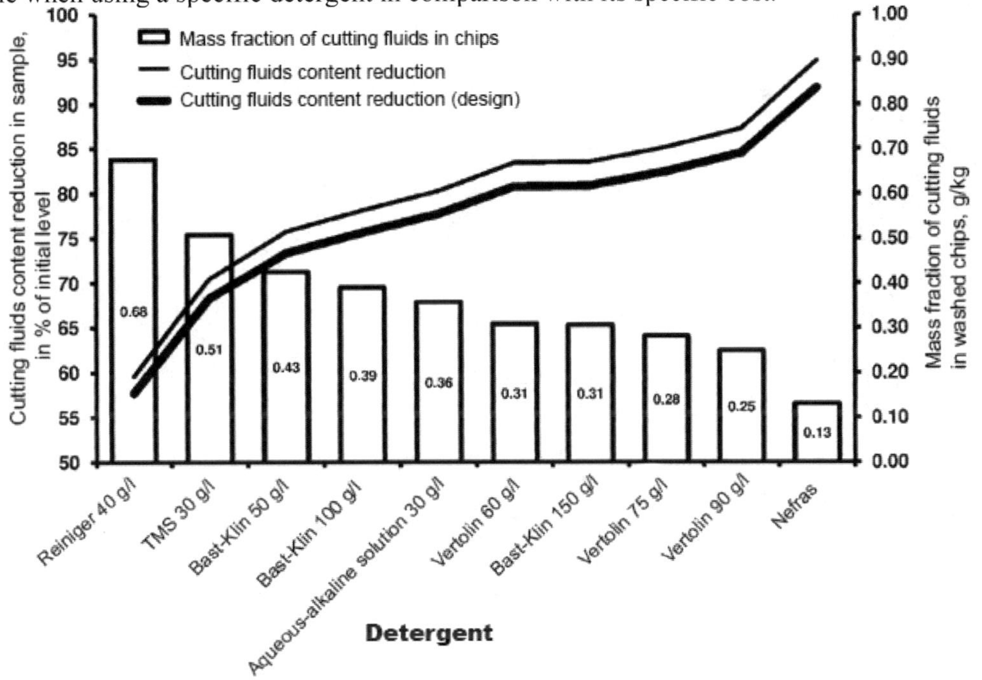

Figure 2. Cutting fluids content reduction curve for various detergents, obtained using the colorimetric express analysis method and corrected based on the gas chromatography data. Calculated mass fraction of cutting fluids depending on the detergent used

The diagram (see Figure 3) shows that a reduction in the cutting fluids content by over 80% is ensured only by the following three detergents: Bast-Klin (with the concentration of 150 g/l), Vertolin (with the concentrations of 60, 75 and 90 g/l), and Nefras.

Considering that Nefras was used in the research as the reference agent and its industrial use is fire hazardous, not environmentally friendly and, as can be seen from Figure 3, the most expensive, Nefras may be excluded from the comparison.

Since the specific cost of Bast-Klin (150 g/l) is more than three times higher than that of Vertolin (even in the highest concentration), the comparison shows that, according to the test results, Vertolin is undoubtedly the best reagent.

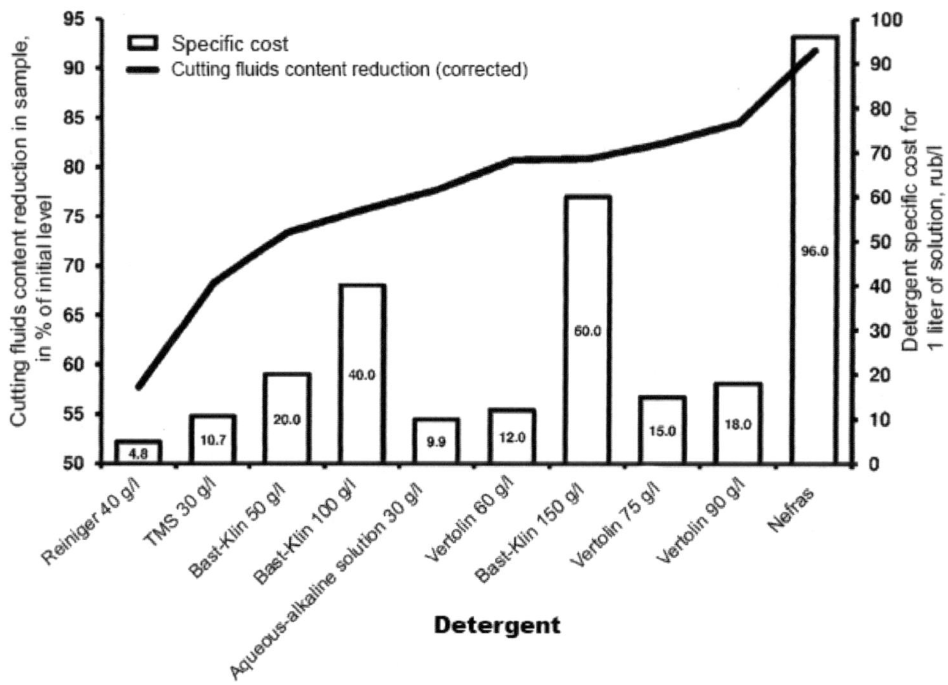

Figure 3. Cutting fluids content reduction in the sample, in % of the initial level, depending on the detergents used, as compared with their specific cost

An increase in the concentration (and cost) of Vertolin by a factor of 1.5 improves the reduction in the cutting fluids content by only 3.8%. Consequently, from an economic point of view, it is best to use Vertolin with the concentration of 60 g/l (with the specific cost of 12 rubles per liter), despite the slightly lower cleaning quality. In addition, lower concentrations of detergents in the solution reduce the environmental impact.

4. Conclusions

- In the course of the research, a colorimetric express method was developed for assessing the contamination of metal chips, which allows quickly and efficiently establishing the quantitative drop in the cutting fluids content in the chips after the washing.

- A reduction of the cutting fluids content by over 80% is ensured only by the following two detergents: Bast-Klin (with the concentration of 150 g/l) and Vertolin (with the concentrations of 60, 75 and 90 g/l). Out of these detergents, Vertolin is deemed optimal, since the specific cost of Bast-Klin (150 g/l) is more than three times higher than that of Vertolin (even in the highest concentration). From an economic point of view, it is best to use Vertolin with the concentration of 60 g/l (with the specific cost of 12 rubles per liter), despite the slightly lower cleaning quality, since an increase in its concentration (and cost) by a factor of 1.5 improves the reduction in the cutting fluids content by as little as 3.8%.

Acknowledgments

The study was carried out under the grant issued by the Russian Science Foundation (project No. 20-79-10125).

References

[1] Asnafi N 2021 Metal Additive Manufacturing – State of the Art 2020 *Metals – Open Access Metallurgy Journal* **11(6)** 867 DOI: 10.3390/met11060867

[2] Ren Xiangyuan, Shao Huiping, Lin Tao and Zheng Hang 2016 3D gel-printing – An additive manufacturing method for producing complex shape parts *Materials & Design* S0264127516304208 DOI: 10.1016/j.matdes.2016.03.152

[3] Kondratyev V B 2017 Global market of rare earth metals *Mining industry* **4(134)** 48–52

[4] Reijonen J 2016 *Utilizing metallic waste streams as raw material for powder-based additive manufacturing* Masters work (Lappeenranta University of Technology) p 131

[5] Simon L, Moraes Carlos A and Modolo R 2016 Recycling of contaminated metallic chip based on eco-efficiency and eco-effectiveness approaches *Journal of Cleaner Production* **153** DOI: 10.1016/j.jclepro.2016.11.058

[6] Shrikant U, Gunjal A, Sudarshan B, Sanap B, Nilesh G and Patil 2020 Role of cutting fluids under minimum quantity lubrication: An experimental investigation of chip thickness *Materials Today: Proceedings* **28(5)** DOI: 10.1016/j.matpr.2020.01.090

[7] Kireinov A V and Esov V B 2017 Current trends in the use of cutting fluids for edge-cutting machining of difficult-to-machine materials *Engineering journal: science and innovations* **2** DOI: 10.18698/2308-6033-2017-2-1591

[8] Yan Lei, Zhang Yunlu, Newkirk Joseph W, Liou Frank, Thomas Eric and Baker Andrew 2018 Investigation of Machining Coolant Residue Cleaning Methods for Ti6Al4V Part Fabrication through Hybrid Manufacturing Process *Manufacturing Letters* DOI: 10.1016/j.mfglet.2018.02.016

[9] Terletskiy S V 2005 On certain problems of metal processing in advanced steel-melting arc furnaces *Casting and metallurgy* **3** 76–78

[10] Tan K A, Mohan Y, Liew K J, Chong S H and Poh P E 2020 Development of an effective cleaning method for metallic parts using microbubbles *Journal of Cleaner Production* **261** 121076 DOI: 10.1016/j.jclepro.2020.121076

[11] Goncharov V S and Tatarinova A D 2020 Development of a metal chip cleaning technology to remove oil contamination for maintenance facilities *Globus* **4(50)** 32–36

[12] Rovin S L, Kalinichenko A S and Rovin L E 2019 Recycling of dispersed metal waste to the production process *Casting and metallurgy* **1** 45–48

[13] Dresel W and Mang T 2017 Lubricants and Lubrication *Removal of Lubricants: Industrial Cleaners* 1089–1124 DOI: 10.1002/9783527645565.ch21

[14] Karmanova S V, Melnikov S A, Shiryaev A A and Nurtdinov A S 2021 Development of a safe and effective parts washing technology with the use of water-based cleaning solutions *Bulletin of PNRPU. Aerospace engineering* **64** 5–15

[15] Starostin A G, Poylov V Z, Karmanova S V, Kotenko A A and Lytkin E V 2020 Removal of process contaminants from aerospace-grade alloy surfaces *Bulletin of PNRPU. Chemical technology and biotechnology* **3** 183–194

IOP Conf. Series: Earth and Environmental Science **938** (2021) 012010 doi:10.1088/1755-1315/938/1/012010

Low-waste innovative production technology for canned quince compote, enriched with natural nutrients

M.M. Rakhmanova[1], M.E. Akhmedov[1], A.F.Demirova[1], Z.A. Abdulkhalikov[1], and N.G. Zagirov[2]

[1]Daghestan State Technical University, 367015, 70, I. Shamyl Ave., Makhachkala, The Republic of Daghestan, Russia,
[2]Federal Research Centre the Subtropical Scientific Centre of the Russian Academy of Sciences, 354002, Russia, Sochi, Yana Fabritiusa Str., 2/28

Abstract. The enterprises processing plant raw materials generate a fairly large amount of waste (secondary products), which can play an important role in solving food, environmental, and energy problems. This work improves the technology for the production of canned quince compote using a two-stage microwave heating of fruits and fruits filled with syrup prepared from infusion obtained from secondary products of cutting and peeling. It has been established that two-stage heating of fruits and microwave treatment of fruits filled with syrup in a glass jar promote an increase in the initial temperature of the product in the jars before sealing and thereby reduces the heat treatment modes. The infusion obtained from the secondary products of the heat-treated fruits provides a significant increase of biologically active components in the finished product. The experimental studies were used to construct a mathematical model for calculating the initial temperature of the product before sterilization, which makes it possible to establish intensive pasteurization modes. New modes of pasteurization of quince compote in various containers have been established. Physicochemical studies confirm the high quality of the finished product in terms of the content of biologically active components.

1. Introduction

The enterprises processing plant raw materials generate a fairly large amount of waste (secondary products), which can play an important role in solving food, environmental, and energy problems. The secondary products can be considered as additional sources of valuable natural substances.

Analysis of literature sources shows that the total amount of agricultural waste reaches 630-650 million tons, and waste from food and processing industries averages 30 million tons per year [1-3]. Recycling provides an expansion of the raw material base for the food industry due to the variety of raw materials of vegetable origin used in the fruit and vegetable industry. The annual volume of waste from the canning industry is 300 thousand tons. The most of waste (up to 70%) is used for feed for farm animals and poultry, and approximately 30% of waste from the fruit and vegetable industry is processed into industrial products.

The global aim is to reduce the amount and involve in the economic turnover (food industry) of secondary raw materials and waste of the agro-industrial complex. The main activities here are optimization of crop and livestock technologies to reduce waste and production losses, development of progressive technological processes for obtaining new types of food products and additives that improve the nutritional and biological value of products, replacement of traditional types of primary raw materials with secondary ones, development and improvement of technologies for the production of high-grade secondary food products, enriched with useful components, etc. [4,5].

Full processing of agricultural raw materials, the most complete extraction of valuable components from it, recycling of production waste will become reserves for increasing production and its efficiency and maintaining ecological natural balance. In this regard, the improvement of processing technologies of secondary raw materials (SRM) and the expansion of the scope of their application are the priority conditions for the use of SRM in industrial food production [6-10].

Waste is also generated in the production of canned compotes from seed raw materials. In particular, in the production of canned quince compote, more than 45% of waste is generated during its cutting and cleaning [11]. So, a key focus area in the technology for the production of canned products with high nutritional value is the search for ways for the complex processing of plant raw materials, which can provide not only savings in material costs but also more complete preservation of biologically active components of raw materials in the finished product.

The work is devoted to the improvement of the technology for the production of canned compote from quince, enriched with natural nutritional food, obtained using the secondary products from cutting and peeling fruits, for cooking syrup. The work aims to develop a technology for the production of canned compote from quince with a relatively high content of vitamins and minerals [12-14], while these components are not introduced from outside, and the enrichment of the finished product is ensured due to their more complete preservation by improving the processing technology itself.

To achieve this goal, the following tasks were solved:

1. Application of effective methods to ensure the safety of components with a high biological value contained in the raw materials;
2. Development of a method for extracting biologically active substances from secondary products obtained by cutting and cleaning raw materials;
3. Development of intensive and thermally soft modes of heat treatment;
4. Evaluation of organoleptic characteristics of canned food prototypes.

2. Materials and methods

The objects of research are quince fruits, secondary products obtained by cutting and peeling quince, and compote from quince. The solids content in the infusion was determined using a refractometer. The product temperature was measured with Chromel-Copel thermocouples connected to a KSP-4 potentiometer.

3. Results and discussion

It is known that the nutritional value of canned products is determined by the presence of biologically active components in them, the content of which is significantly reduced during their processing and mainly depends on the level of technical perfection.

A comprehensive analysis of the traditional technology for the production of canned compote from quince shows the presence of several well-known drawbacks, which, in general, affect the vitamin content and mineralization of the finished product. To accomplish the assigned tasks, it is proposed to use waste (secondary products) obtained during cutting and cleaning fruits for the preparation of syrup. Instead of blanching fruits in water, it is proposed to carry out microwave blanching in the jars, after packaging the fruits and the use of gentle sterilization regimes. To increase the yield of dry substances from secondary products, they are exposed to microwave field for 60-90 seconds, followed by grinding and extraction in warm water at 60 °C for 10 -15 minutes.

The infusion obtained from secondary products after cleaning and cutting the fruit is used to prepare the syrup, and it contains not only carbohydrates, which saves more than 10 kg of sugar, but also a sufficient amount of vitamins and microelements. Another element of the proposed technological scheme is the use of microwave blanching of fruits packaged in glass jars for 60-120 seconds, depending on the volume of the glass jar. Microwave blanching contributes not only to ensuring the results achieved by blanching in hot water, but also contributes to an increase in the nutritional value of the compote, preventing the loss of vitamins and microelements, at the same time

IOP Conf. Series: Earth and Environmental Science **938** (2021) 012010 doi:10.1088/1755-1315/938/1/012010

provides an increase in the temperature level of the product entering for sterilization and saving more than 50 thousand kJ of heat per 1,000 standard cans of product.

The quality of canned products in a hermetically sealed container, subjected to heat treatment to suppress the vital activity of microorganisms, largely depends on the duration of the heat treatment modes and the degree of uniformity of heating the product throughout the entire volume of the glass jar [12-14].

Therefore, the search for methods that reduce the duration of pasteurization and preserve the vitamin composition of finished products is an important direction in improving the technological processes for the production of food rich in biologically active components.

Analysis and evaluation of methods that reduce the pasteurization time show that the most optimal is to increase the temperature level of the semi-finished product before the start of pasteurization. Microwave blanching also helps to preliminarily reduce the microbiological charge of the semi-finished product used for pasteurization [12, 14].

New technical solutions for preliminary preparation increase the temperature of the semi-finished product before sterilization by more than 42 °C, which makes it possible to use high-temperature heat carriers and accelerated modes of thermal sterilization.

Table 1 shows characteristics of a semi-finished product in glass jars exposed to microwave fields of various parameters.

Table 1. Data on microwave treatment of the semi-finished product of the quince compote

Glass jar volume, l	Microwave field power, W	Treatment time, s	Product temperature, °C	
			Treated by the microwave field	Obtained by the traditional technology
1.0	700	60	63	52
		120	72	51
		180	82	50
1.0	420	60	60	50
		120	65	51
		180	70	52
1.0	280	60	55	51
		120	58	51
		180	62	50
0.5	700	60	70	51
		120	84	51
		180	93	51
0.5	420	60	62	51
		120	68	51
		180	74	51
0.5	280	60	58	51
		120	63	51
		180	71	51
0.65	700	60	67	52
		120	75	50
		180	89	51
0.65	420	60	58	51
		120	65	51
		180	70	51
0.65	280	60	53	51
		120	58	51
		180	64	51

Using mathematical processing of the experimental data we obtained an equation for determining the temperature of a semi-finished product in a glass jar before sealing (T, °C) with a two-stage exposure to a microwave field. The following main parameters were taken into account: the glass jar volume (V, l), the power of the microwave field (P, kW) and the exposure time of the microwave field (t, s):

$$T = 57.2 - 11.29V - 10.29P - 0.06t + 28.85VP + 0.55Vt + 0.08Pt \quad (1)$$

The intensive modes of pasteurization can be established using the calculated temperature of the product before sealing and the obtained equation (1) [15-17].

Changes in the parameters of temperature and microbial lethality during pasteurization of quince compote in a jar of SKO 1-82-500 according to the new regime $80.\frac{5 \quad 15 \quad 20}{90 \quad 100 \quad 40} 88$ kPa are shown in Figure 1. Here 80 is the initial temperature of the product before sterilization, °C; 5 is the duration of the temperature rise in the autoclave up to 100 °C, min; 15 is the duration of sterilization, min; 20 is the duration of water cooling in the autoclave, min; 90 is the initial temperature of water in the autoclave, °C; 100 is the sterilization temperature, °C; 40 is the final temperature of water in the autoclave, °C.

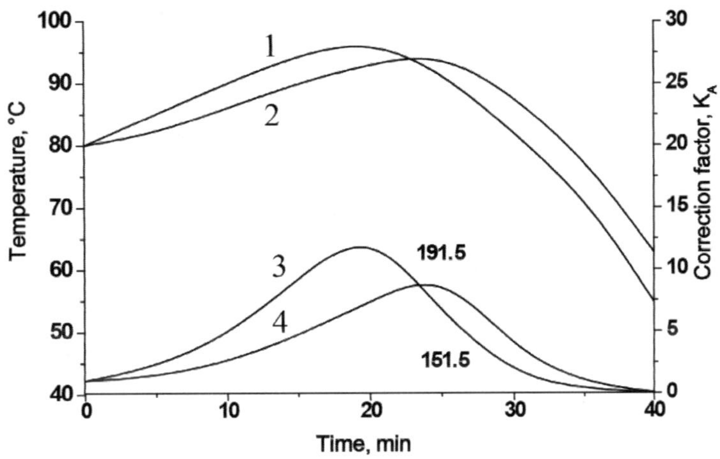

Figure 1. Graphs of changes in temperature parameters (1,2) and death of microorganisms (3,4) in the parietal (1,3) and middle (2,4) areas of a 0.5 l glass jar during pasteurization of quince compote using the developed method

Analysis of the graphs shows that the developed mode achieves the required sterility, which is confirmed by the sterilizing outcomes equal to 191.5 and 151.5 conventional minutes, respectively, and at the same time, the total pasteurization time is reduced by 20 minutes.

The same studies were carried out for compote in jars of 0.35 and 1.0 l. Based on the results of the studies performed, new modes of pasteurization of quince compote were established (Table 1).

Table 1. Intensive modes of pasteurization of quince compote

No.	Jar volume, l	Pasteurization modes according to the existing technology	Modes of pasteurization by heating fruits in a microwave field
1	0.35	$\frac{20 \quad 15 \quad 20}{100}$ 118 kPa	$80.\frac{5 \quad 12 \quad 20}{90 \quad 100 \quad 40}$ 88 kPa
2	0.5	$\frac{20 \quad 25 \quad 20}{100}$ 118 kPa	$80.\frac{5 \quad 15 \quad 20}{90 \quad 100 \quad 40}$ 88 kPa
3	1.0	$\frac{25 \quad 35 \quad 25}{100}$ 118 kPa	$80.\frac{10 \quad 25 \quad 25}{90 \quad 100 \quad 40}$ 88 kPa

An improved technological scheme for the production of quince compote using the proposed technical solutions is shown in Figure 2.

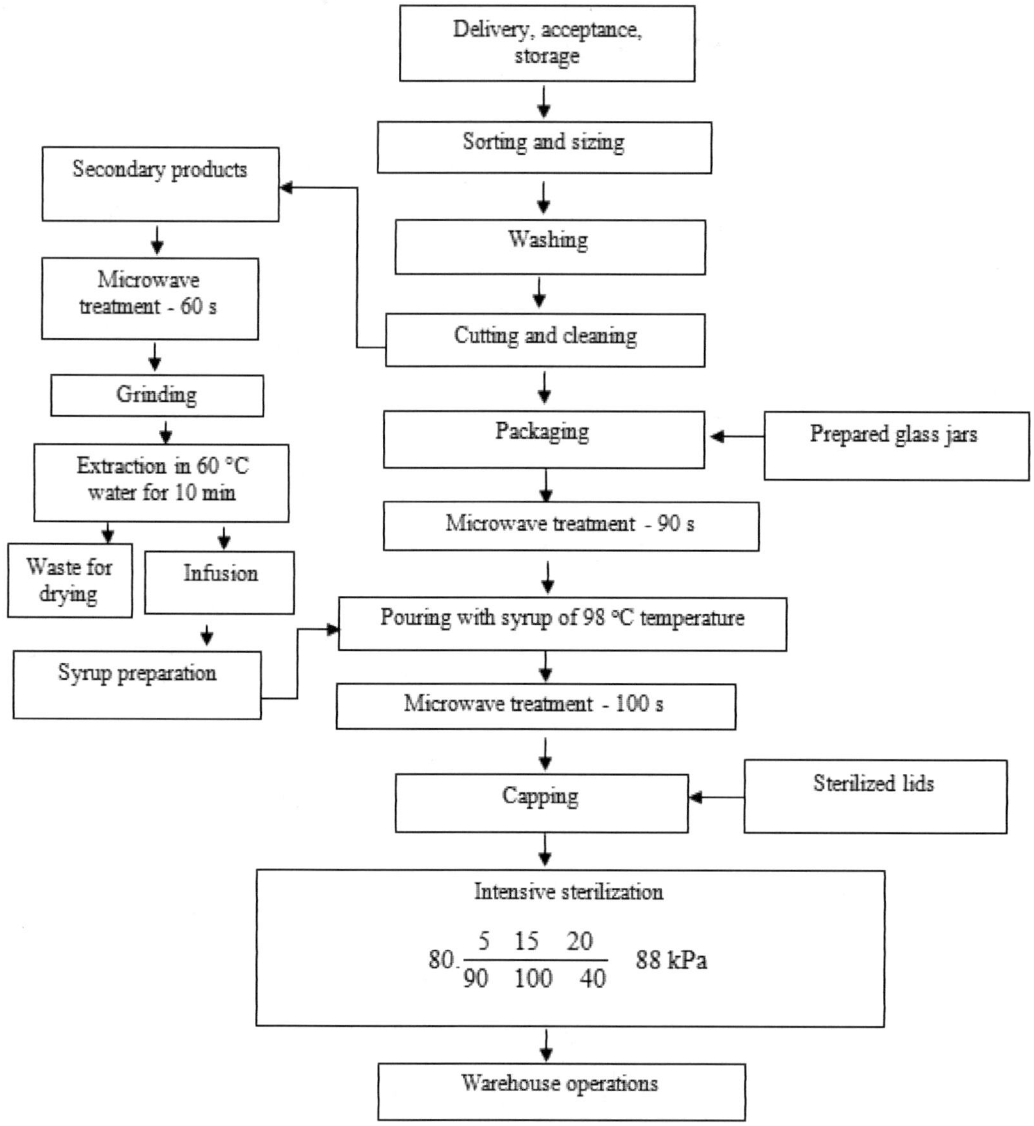

Figure 2. Low-waste innovative technology for quince compote in a 0.5 l glass jar using secondary resources and gentle heat treatment

The data on the presence of vitamin C in quince fruits and compote prepared using different technologies are shown in Figure 3.

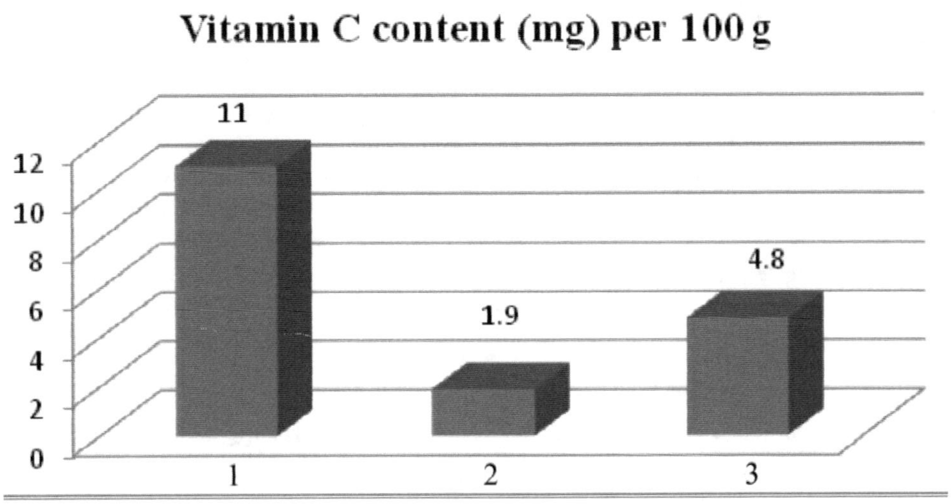

Figure 3. The content of vitamin C in the raw materials and compote from quince, depending on the production technology: 1 - raw materials; 2 - traditional technology; 3 - improved technology

This technology for the production of canned quince compote was implemented at OOO "Agropromyshlennyy kombinat. Nagornyy Daghestan" of the Republic of Daghestan. Production tests have confirmed the high quality of the products obtained by the developed technology.

4. Conclusions

Analysis of the research results confirms the feasibility of the new technology, as ensuring the maximum preservation of the nutrient composition of the raw materials in canned compote. It can be used in the canning industry.

References

[1] Rodionova L Ya 2004 Technological and experimental substantiation of the technology of pectin-containing products for functional purposes: thesis for the degree of dr. of tech. sciences (Krasnodar) p 470

[2] Flaumenbaum B L 1982 Basics of food preservation (Moscow: Light and food industry) p 272

[3] Federal classification catalog of waste Available at: http://www.fkko.ru

[4] Mikryukova N V 2012 The main aspects of obtaining functional food products Young Scientist **12** 90–92

[5] Mkhitaryan G A, Lesnov A P and Tkachenko V M 2009 Modern technologies for processing beet pulp Sugar beet **2** 33–35

[6] Gorbachev M G 2013 Development of the technology of specialized powdered food products for the correction of protein and energy deficit: thesis for the degree of cand. of tech. sciences (Moscow) p 196

[7] Koshchaev A G, Plutakhin G A, Fisenko G V and Petenko A I 2008 Waste-free processing of sunflower meal Storage and processing of agricultural raw materials **3** 66–68

[8] Woollen A 1990 Functional foods – a new market? Food Rev. **17(4)** 63–64

[9] Louise Wicker, Yookyung Kim, Mi-Ja Kim, Brittnee Thirkield, Zhuangsheng Lin and Jiyoung Jung 2014 Pectin as a bioactive polysaccharide – Extracting tailored function from less Food Hydrocolloids **42(2)** 251–259

[10] Golubev I G, Shvanskaya I A, Konovalenko L Yu and Lopatnikov M V 2011 Waste recycling in the agro-industrial complex: a reference book (Moscow: FGBNU

"Rosinformagrotech") p 296
[11] Collection of technological instructions for the production of canned food 1977 volume 2 (Moscow: Food industry)
[12] Akhmetov M, Demirova A, Piniaskin V and Rakhmanova R A 2020 New technological and technical solutions in dietary pear compote production E3S Web of Conferences **161**
[13] Akhmetov M, Demirova A, Abdulkhalikov Z, Daudova T and Daudova L 2020 An enhanced technology of pear compote production through direct blanching with sugar syrup in glass jars and a device for its implementation E3S Web of Conferences **161**
[14] Akhmedov M E, Gammatsaev K R, Omarova A A and Rakhmanova R A 2018 New modes of thermal sterilization of quince compote in SKO 1-82-500 jar Collection of scientific works of teachers, staff, graduate students and students of the technological faculty of DSTU 30–32
[15] Babarin V P 2006 Sterilization of canned food (SPb: GIORD) p 312
[16] Davis C O, Smith O and Olander J 1965 Microwave processing of potato chips Pt. J. Potato Chippe **35** 38–58
[17] Kasyanov G I, Demirova A F and Akhmedov M E 2014 Innovative technology of sterilization of fruit and vegetable raw materials Reports of the Russian Academy of Agricultural Sciences **6** 57–59

ASSESSMENT OF THE IMPACT OF VEHICLE EMISSIONS ON THE ATMOSPHERE OF THE CITY OF VORONEZH AND THEIR MINIMIZATION

V. Ya. Manokhin[1] and M. V. Manokhin[2]

[1] Voronezh State Technical University
[2] Voronezh State Technical University

Abstract. The paper considers the emissions of pollutants from road transport, in the total air pollution they make up 70 % on the example of a particular metropolis. The highway is considered as a linear source consisting of single point sources of pollution. Special attention is paid to suspended fine particles in the solid phase. It is determined that when assessing the level of environmental pollution, it is important to take into account not only the total number of vehicles, but also the ratio of cars of different categories, since they, having different emission factors, supply unequal volumes (mass) exhaust gas components. The variability of the traffic flow density on the streets of six administrative districts of Voronezh and the complexity of zoning the city territory because of three-year monitoring were established. It is proposed to place accounting points on streets that differ in their status in the general transport network of the city in order to obtain reliable data during the monitoring period [11]. The simulation of diffusion movements and migrations of solid-phase air pollutants of roadside territories is carried out. The model of diffusive displacements and migrations of heavy metals is considered in relation only to passive contaminants. The choice is implemented and the method of studying the level of dustiness of motor transport roads is recommended, which allows fixing dust particles from 0.1 microns to 250 microns in the air. An algorithm of an improved system for monitoring the content of fine dust in the atmosphere of a residential zone is proposed. [16]. To minimize the emission of suspended substances, a two-stage emission purification system consisting of a gas-liquid absorber and a catalytic adsorber is proposed.

1. Introduction

The problems of protecting the atmosphere of megacities are complicated by the prevalence of motor vehicle emissions in the total atmospheric pollution. The native city, the metropolis of Voronezh, with a population of 1 million inhabitants, was chosen as the object of research.

The problem of air pollution is particularly acute in megacities, where the amount of transport emissions significantly reduces the quality of life of the population.

The purpose of the research is to assess the contribution of road transport in the gross emission of air pollution at a specific object with a developed industrial structure of the technosphere and to develop solutions for their reduction and subsequent minimization. [3]

Statistical data on gross emissions of polluting components into the atmosphere from stationary and mobile sources from open bases from the Russian Natural Supervision of the Voronezh Region are used for the research. The «Statistics» software packages are used for data processing.

Now, there is no single methodology that would allow us to assess the dispersed composition of dust in the air of residential areas in Russia does not exist. To determine the dispersed composition of

IOP Conf. Series: Earth and Environmental Science **938** (2021) 012011 doi:10.1088/1755-1315/938/1/012011

dust in the residential area, the «Method of microscopic analysis of the dispersion composition of dust using a personal computer (PC)» was used. This method makes it possible to determine the presence of particles in the air from 0.1 microns to 250 microns, which is suitable for this type of pollution. Mathematical zoning of the diffusion process of roadside pollution by solid particles is implemented using software (Ecologist software, OND-86, StatSoft STATISTICA program for statistical data processing, ERSI ArcGIS for processing geographical information). [5]

2. Experimental and analytical studies on emissions into the air zone of the Voronezh megalopolis

The analysis of statistical data shows that due to the growing number of individual transport, the number of transport emissions into the atmosphere of cities is increasing. At a specific object of the technosphere – the megalopolis of Voronezh, after statistical processing of information indicators on air emissions, it was found that the number of gross emissions from stationary and mobile point single sources of pollution increases annually (thousand tons/year). The results of processing using the StatSoft STATISTICS program are shown in Figure 1.

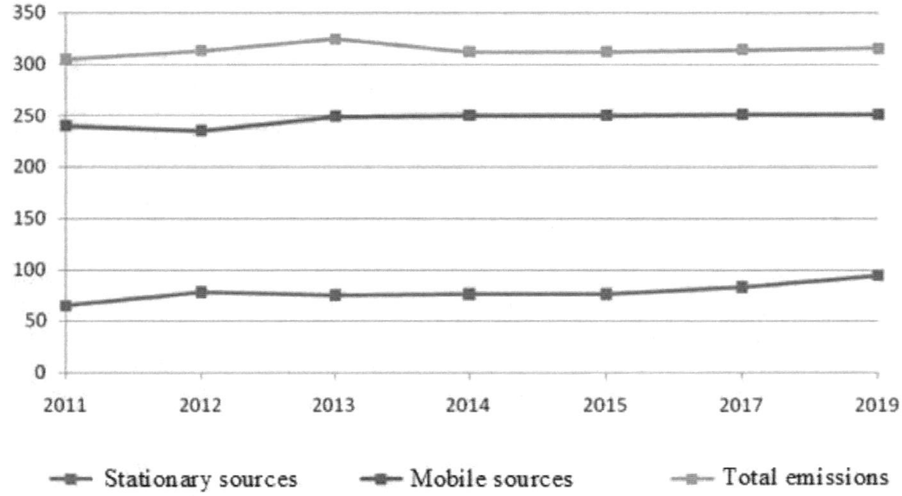

Figure 1. Emissions from mobile and stationary sources of atmospheric pollution in Voronezh

Further, the assessment of the contribution of road transport in the gross emission for air pollution is implemented on the specific example of the city of Voronezh with the use of similar information technologies. The following pattern is revealed. In Voronezh, 70% of the total emissions are carried out by motor transport. Table 1 shows the information after processing statistical data on the number of cars per 1000 inhabitants for various cities of the Russian Federation for comparison with the city of Voronezh. It is established that the Voronezh metropolis is one of the leaders in this indicator and takes the third place, the data are presented from 2011 to 2018. In 2019, this indicator reached the level of 330 units. [1]

Table 1. The number of cars in the megacities of Russia

№	City	Park, thousand pcs.	Population, thousand people.	Security per thousand people, pcs.
1	Samara	390.9	1170.0	334
2	Saint Petersburg	1665.3	5225.7	319
3	Voronezh	330.4	1054.111	319
4	Moscow	3782.8	12330.1	307
5	Kazan	368.5	1217.0	303
6	Yekaterinburg	446.5	1477.7	302
7	Krasnoyarsk	312.5	1067.9	293
8	Rostov-on-Don	319.2	1119.9	285
9	Ufa	311.6	1121.4	278
10	Nizhny Novgorod	352.0	1275.5	276
11	Novosibirsk	433.1	1584.1	273
12	Omsk	321.3	1178.1	273
13	Chelyabinsk	320.4	1192.0	269
14	Volgograd	256.3	1016.1	252
15	Perm	246.5	1041.9	237

We will analyze the dynamics of emissions of pollutants into the atmosphere in the Voronezh region from stationary and mobile sources (thousand tons / year). The results of the research on the processing of statistical data are presented in Fig. 2.

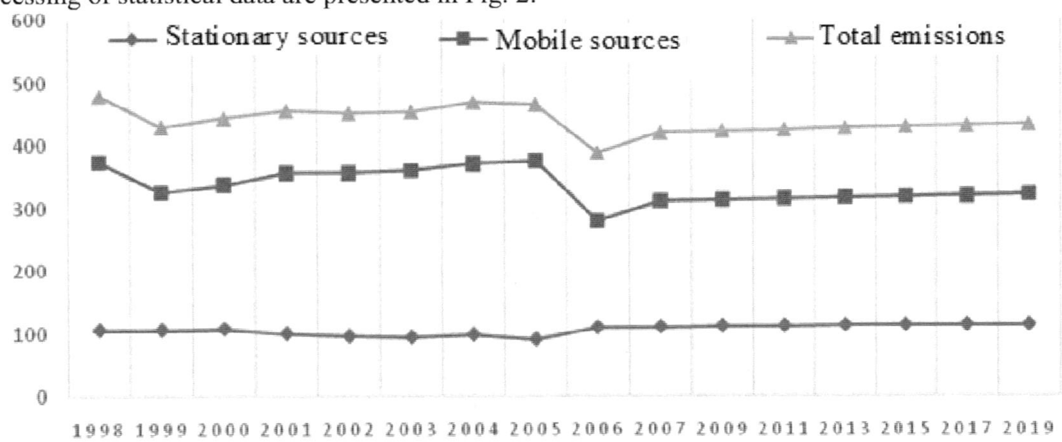

Figure 2. Dynamics of emissions of pollutants into the atmosphere of the Voronezh region from stationary and mobile sources (thousand tons / year)

Based on the analysis of statistical information, it was revealed that gross emissions from stationary and mobile sources tend to increase compared to 2006. According to Fig. 2, the dependence of the amount of gross emissions from mobile sources (thousand tons / year) on the year from 1998 to 2019 can be divided conditionally into two components: 1998-2006 and 2006-2019. Obviously, a number of socio-economic factors caused the year 2006, but this analysis of the reasons was not included in the task of the study. The amount of emissions from stationary sources from 1998 to 2019 fluctuates at the same level of about 100-120 thousand tons / year, the reason for this, as we see one - the number of exact objects of the technosphere remains approximately at the same level. [2,4]

3. Zoning of the diffusion process of roadside pollution by solid particles

The object of research was motor vehicles of five categories of cars: passenger cars, trucks (with a load capacity of up to 3 tons, 3-5 tons, 8 or more tons), buses. The observation points were located on the streets of six administrative districts of Voronezh. During the observation period in the Central District, the traffic load was determined in 25 points on 19 streets. In the Sovetsky district – in 10 points on 9 streets. In the Kominternovsky district – in 34 points on 19 streets. In the Railway district – in 7 points on 7 streets. In the Levoberezhny district – in 15 points on 9 streets. In the Leninsky district – in 3 points on 3 streets. The location of the accounting points and their number changed depending on the capabilities of the observers.

The materials obtained indicate that the traffic load on the streets of the city varies by year, both within individual districts and in the city as a whole. Although the number and specific bindings of accounting points to the transport network of the city were mostly different in different years, the average indicators for the district can be considered acceptable. An attempt to zone the territory of the city (Fig. 3) on the basis of the results obtained showed that this is impossible at this stage of work. Only in one case – in the Sovetsky district – for two years of research and an average of three years, similar results were obtained, determining its rank position by the number of vehicles registered on the streets per day. [4,6]

Table 2. Distribution of motor transport flows in the administrative districts of Voronezh

City district	Number of vehicles per day passing through the accounting points by year	
	2018	2019
Central	9183	15122
Soviet	8683	13342
Kominternovsky	13772	13756
Levoberezhny	6154	20805
Railway	23974	13213
Leninsky	9883	32694
Average	11941	18155

When assessing the level of environmental pollution, it is important to take into account not only the total number of vehicles, but also the ratio of cars of different categories, since they, having different emission factors, supply unequal volumes (mass) exhaust gas components. Considering our data in this aspect, we noted that for this indicator there is a noticeable difference between the districts of the city, reflected in the form of two contrasting examples in Fig. 4, 5.

The general fact is that there is a clear predominance of the number of passenger cars, which, for most components, give the greatest mass of emissions in exhaust gases. Within the district, on different streets, the composition of the automobile flow and the mass of emissions of seven exhaust gas components (CO, CO_2, NO_x, SO_2, C_mH_n, Pb and solid particles) also varies, as, for example, shown in Fig. 4, 5 for several observation points at five intersections and streets in the Central district of the city. [10,7,8]

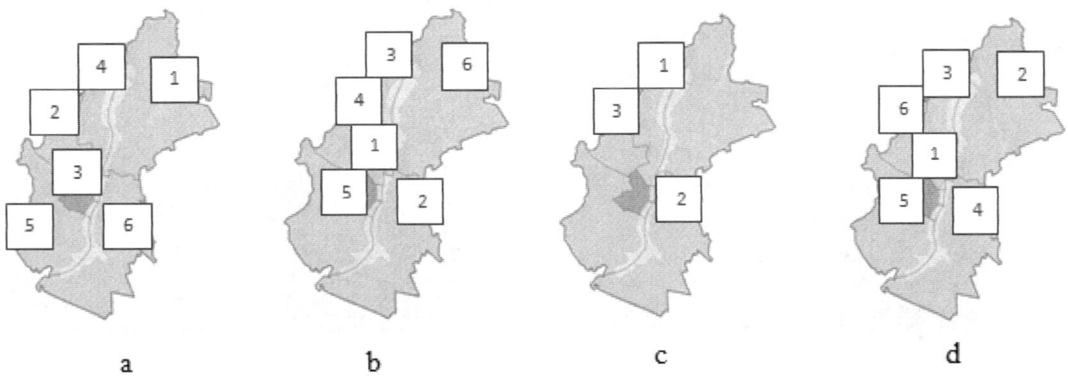

Figure 3. Zoning of the city territory based on materials taken into account in different years (from left to right – a-2017, b-2018, c-2019, d-average for three years)

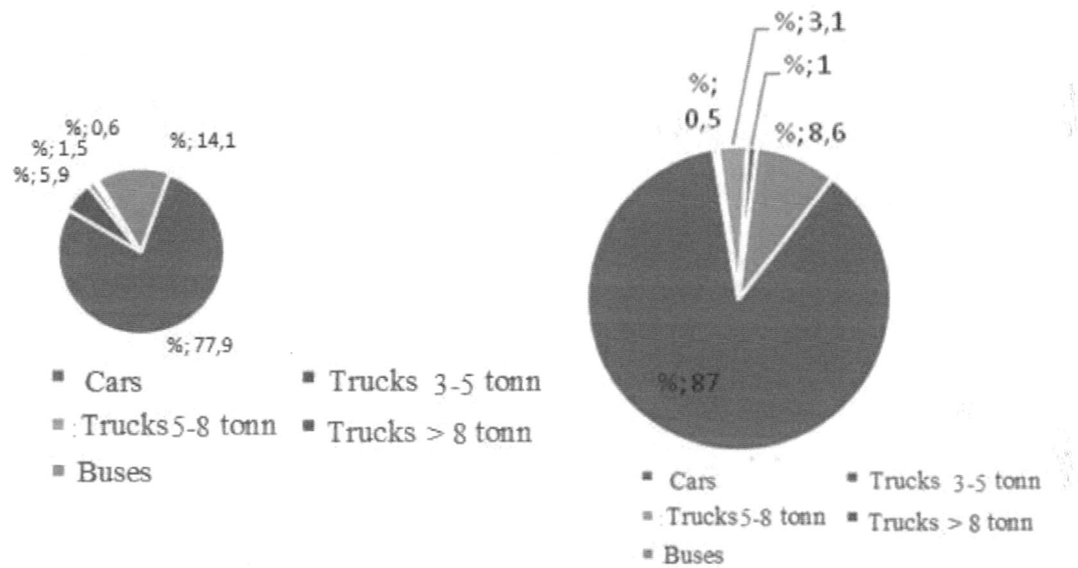

Figure 4. The composition of the traffic flow (units/day) in the Central district of the city (left-2018, right-2019)

IOP Conf. Series: Earth and Environmental Science **938** (2021) 012011 doi:10.1088/1755-1315/938/1/012011

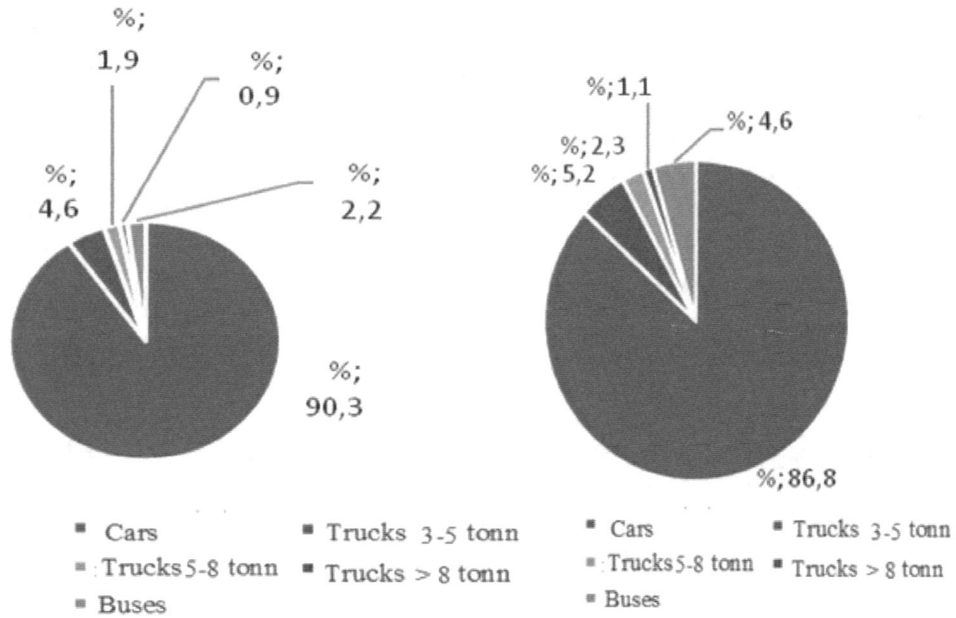

Figure 5. The composition of the traffic flow (units/day) in the Railway district (left - 2018, right-2019)

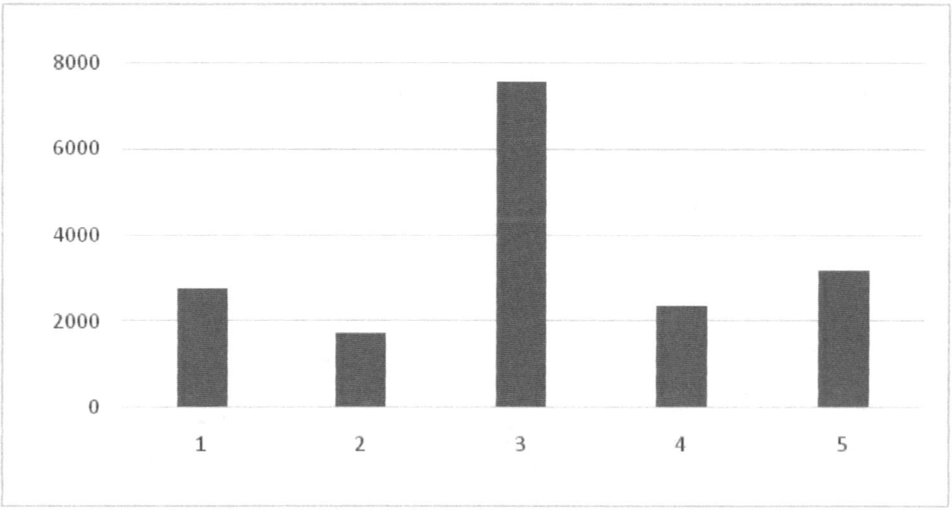

Figure 6. The ratio of the number of cars (units/day) of five categories on the streets of the Central District of Voronezh (2018). Designations: along the abscissa axis of the street numbers: 1 - the intersection of Uritsky Street, etc. Truda; 2 -the intersection of Fr. Engels and Krasnoarmeyskaya streets; 3 – intersection of Truda Ave. and Moskovsky Ave.; 4 – Transportnaya str.; 5-intersection of Komissarzhevskaya Str. and Fr. Engels

Therefore, the available results are of a preliminary nature, they are the initial stage of the project development, which assumes the expansion and deepening of research in subsequent years. It is necessary to note the methodological aspects of solving this problem. To obtain more reliable data throughout the entire monitoring period, the accounting points should be located on streets that differ in their status in the general transport network of the city. At the same time, the localization of accounting points in different years should be as close as possible.

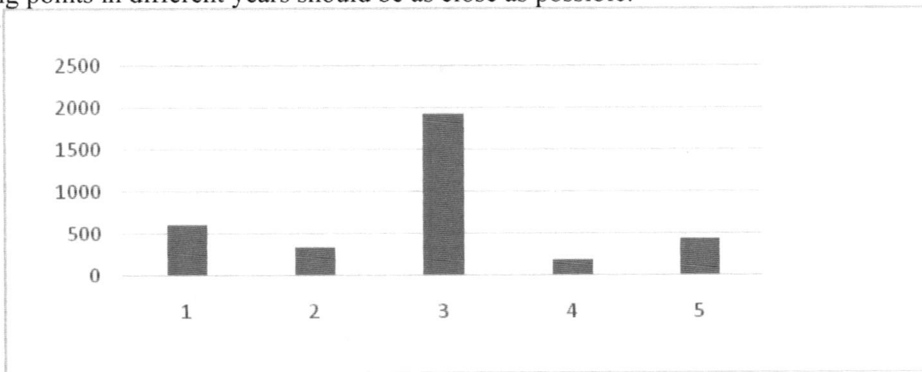

Figure 7. The ratio of mass emissions of seven components of car exhaust gases on the streets of the Central District of Voronezh (2019). Designations: along the abscissa axis, street numbers: 1-the intersection of Uritsky Street, etc. Truda; 2-intersection of Fr. Engels street and Krasnoarmeyskaya; 3 – the intersection of Truda Ave. and Moskovsky Ave.; 4-Transportnaya Str.; 5 – the intersection of Komissarzhevskaya Str. and Fr. Engels

In the mathematical model, the highway is considered as a linear source of pollution (a set of distributed single sources). Table 3 illustrates the results of calculating specific emissions on a two-kilometer section of the Voronezh city highway, depending on the mileage indicator of passenger transport. [10,7,8]

Table 3-Mileage emissions of pollutants

Name of the group of cars	Group number	Emission of a polluting substance, g / sec						
		CO	NOx (converted to NO_2)	CH	Soot	SO_2	Formaldehyde	Benzpyrene
Passenger cars	I	0.2338	0.087	0.0696	0.00145	0.00174	0.000406	0.000696

The problem of accounting for the emission of suspended substances when driving vehicles is determined by the presence of organized emissions from internal combustion engines (ICE). Basic pollutants: soot, metal oxides and unorganized emissions coming from brakes, clutches, when tires and road surfaces wear out, as well as air flows arising from the movement of the car. The obtained data on mileage emissions were used for further mathematical modeling.

The model of diffusion displacements and migrations of heavy metals is based on the concept: in view of the multifactorial nature of the problem, it should be considered in relation to passive priority contaminants. The simulation is implemented with a restriction: the polluting ingredients do not undergo modifications as a result of chemical reactions with water vapor, ozone and other components before falling on the underlying soil surface. The absence of annihilation processes with polluting components is accepted.

The next accepted condition. Stokes forces affect the solid particle, in addition, the vertical component of turbulent diffusion contributes to the deposition of particles [20].

These factors are taken into account in the mathematical zoning of the diffusion process of pollution of roadside territories with solid particles.

The diffusion-convective mechanism of pollutants movement is described by the standard diffusion equation with a mass source g1. The source of the emission is the exhaust manifold of the engine. The formulation of the diffusion problem is based on the following independent submodels: the movement of an instantaneous source in a stationary medium, the movement of the medium around a stationary source, or the simultaneous movement of the medium and the source.

The solution of the diffusion problem is obtained using the potential theory, which has been successfully applied in the theory of thermal conductivity [18]. For light aerosols in [19], the Green function was used to solve the diffusion equations and solutions describing the migration of contaminants without taking into account the conditions of their deposition were obtained.

We will consider: that the process of spreading pollutants in the air occurs so quickly that the pollutants do not annihilate. This is confirmed by the fact that the decomposition of the amount of a substance expressed in moles, for example, nitrogen oxides, occurs in a few days, and its movement takes place within a shorter time. Solid contaminants are deposited near the roadbed almost immediately. If we consider a quasi-stationary process of moving aerosols of pollutants, then instead of a point source, we can take a linear one with an average emission density g1 (in time):

$$g_1 = 1/\mathrm{T}\int_0^T g_1 d1 \tag{1}$$

where T – averaging period, $dc_1/dt \approx 0$.

To assess transport pollution, the road sections where the traffic intensity is the highest are considered, including the dust content in urban areas near the busiest sections of road and street networks [4].

The dependences of the content of PM_{10} (particles of ingredients and suspended particles with a diameter of 10 micrometers (microns) or less) in the city's atmosphere on wind speed and relative humidity are established (Fig. 8,9). To fully assess the influence of all these factors on the concentrations of PM_{10} and suspended particles in the air of the studied intersections, data were processed using regression and correlation analyses [5].

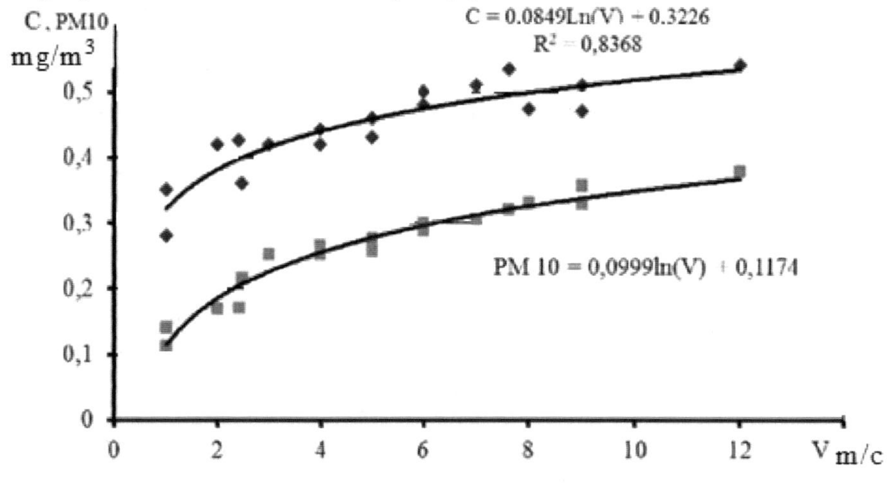

Figure 8. Dependence of PM_{10} and dust concentration on wind speed

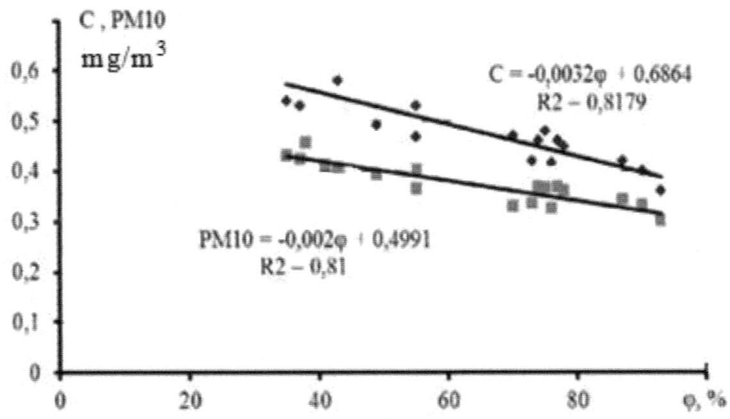

Figure 9. Dependence of PM_{10} and dust concentration on air humidity

At the next stage, the methods that implement the process of determining the dispersed composition of dust are analyzed. Most of them are used to study the level of dustiness of motor transport roads.

Now, there is no single methodology that would allow us to assess the dispersed composition of dust in the air of residential areas in Russia does not exist. To diagnose the dispersed composition of dust in the residential area, the "Method of microscopic analysis of the dispersion composition of dust using a personal computer (PC)"was used. The chosen method makes it possible to determine the presence of particles in the air from 0.1 microns to 250 microns, which is suitable for this type of pollution.

The description of the dispersed composition and the diagnosis of the dust concentration with was implemented based on the theory of random functions. One of the main features of the movement of vehicles in the city is its unsteadiness, which is associated with the uneven speed of movement, heterogeneity of the composition of the traffic flow, changing traffic intensity and other factors.

In order to calculate the average number of dust concentration output for a fixed level (standard value), the Rice formula is used.

To determine the average concentrations of suspended substances exceeding the standard levels in the city's atmosphere, based on the results of the conducted studies, daily maxima were derived throughout the day and a sample (volume = 70) of daily maximum dust concentrations was obtained.

To improve the monitoring system for the content of fine dust in the atmosphere of the residential zone, the following algorithm of functioning is proposed, shown in Fig. 10.

Figure 11 shows the first stage of cleaning of internal combustion engine emissions, which is a gas-liquid absorber. The absorption process proceeds with the release of heat, which often has to be removed from the absorber with the help of circulating irrigation, cooled in refrigerators. The technological scheme of absorption with preliminary purification from gasoline components (extraction of butanes and pentanes) of the raw gas and saturation of the absorbent with methane and ethane is shown in Fig. 11.

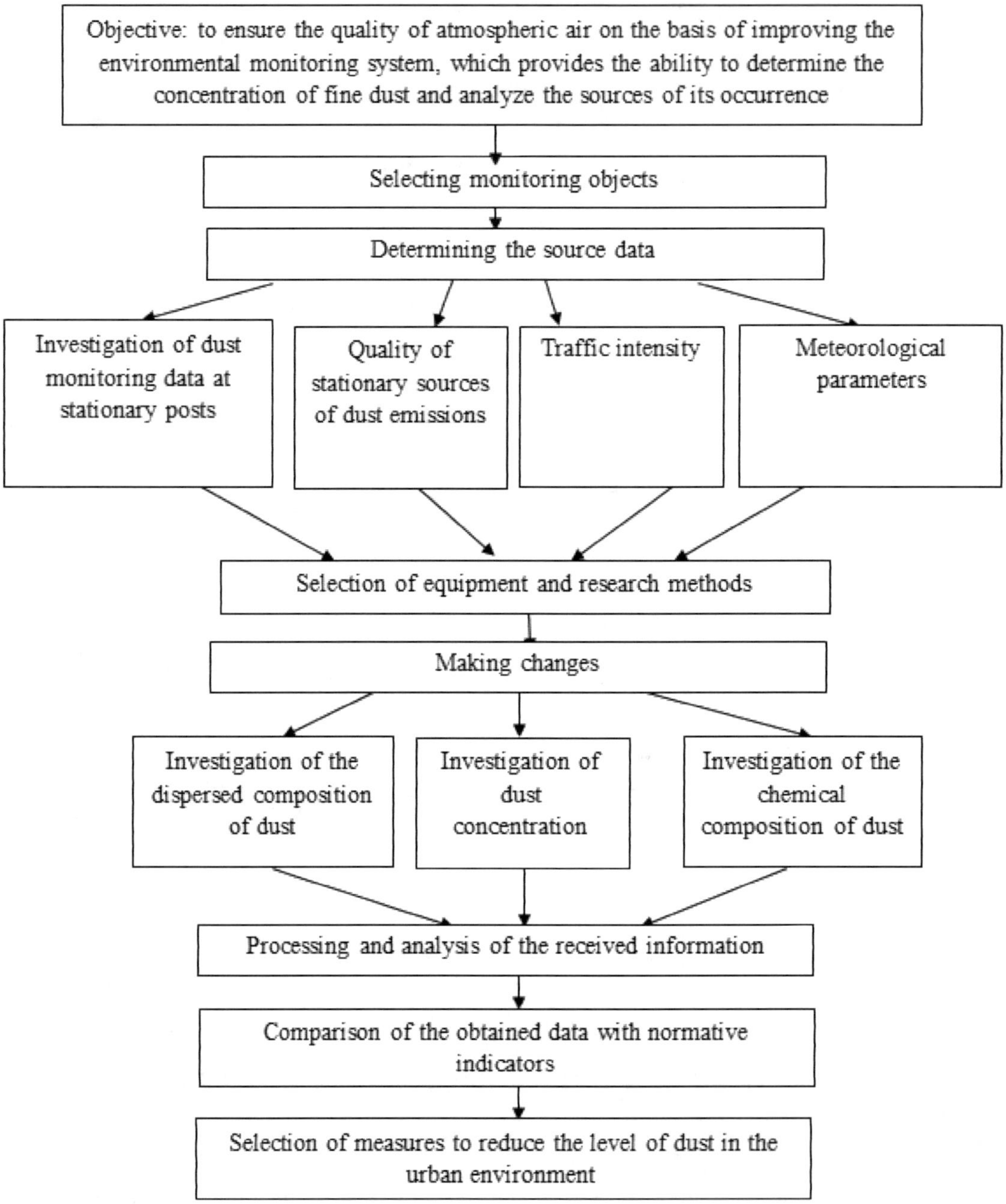

Figure 10. Scheme of organization of monitoring of dust content of atmospheric air

The second stage of emission purification is a standard catalytic adsorber (Fig. 12). A vertical cylindrical adsorber of periodic action is presented. A horizontal support and distribution grid is located in the body of the device, on which a granular layer of adsorbent is placed. The initial mixture passes through the layer from top to bottom, while the adsorbed component-the adsorbent-is absorbed by the adsorbent. The mixture purified from the adsorbent (treated) passes through the support grid and leaves the device. To extract the absorbed substance, the adsorbent layer is treated with water vapor, passing it through the layer from bottom to top. Part of the steam condenses, the resulting condensate is discharged through the lower fitting of the device. The rest of the steam with the adsorbent contained in it leaves the apparatus from above. Using this device allows you to reduce the amount of pollution in the atmosphere by 90 %. The calculated averted damage caused to the atmosphere of the city of Voronezh will amount to more than 6 billion rubles per year. In order to reduce the amount of emissions from the internal combustion engines of cars, a two-stage emission purification system is proposed.

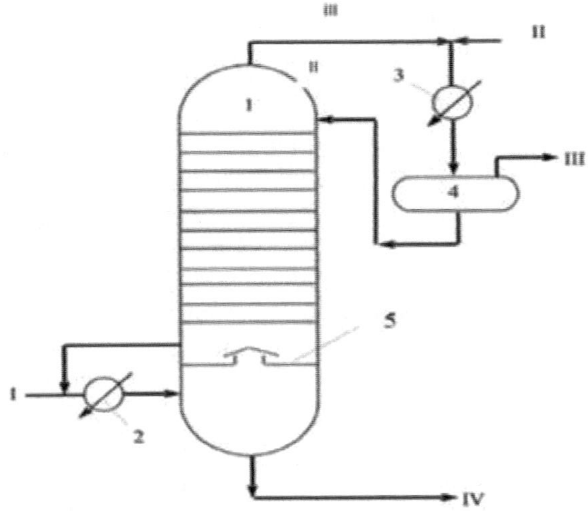

Figure 11. General view of the gas-liquid absorber – 1 stage cleaning (1-Absorber, 2,3-refrigerators, 4-container, 5-semi-dry absorber plate, I-raw gas, II-lean absorbent, III – distilled gas, IV-saturated absorbent)

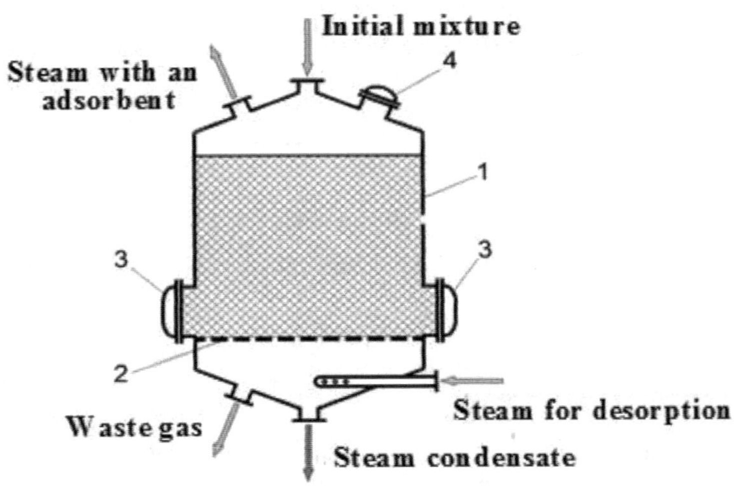

Figure 12. Vertical cylindrical adsorbent with a fixed adsorbent layer: 1-housing, 2-support and distribution grid, 3-hatches for unloading adsorbent, 4-hatch for loading adsorbent – 2 cleaning stage

Based on the results of the research, the following conclusions can be drawn:

1. Emissions from motor transport into the atmosphere of the city of Voronezh exceed 70% of the total (taking into account emissions from stationary sources).

2. The number of motor vehicles per thousand people has reached 330 units. Voronezh ranks 3rd in this indicator in Russia.

3. The emissions of suspended substances from internal combustion engines, clutches, brakes, tire wear and road surface are considered in detail. Special attention is paid to suspended microcrystalline dust particles.

4. It is necessary to note the proposed methodological aspects of solving the problem of zoning the territory of megacities of the technosphere. It is proposed that in order to obtain reliable data throughout the entire monitoring period, accounting points should be located on streets that differ in their status in the general transport network of the city. Localization of accounting points in different years should be as close as possible.

5. On the basis of three-year monitoring of the traffic density on the streets of six administrative districts of Voronezh, the variability of this indicator and the complexity of zoning the territory of the city were established.

6. Modeling of diffusion movements and migrations of solid-phase air pollutants of roadside territories is carried out.

7. The model of diffusion displacements and migrations of heavy metals is considered in relation only to passive contaminants. It is established that the considered passive polluting components do not undergo modifications as a result of chemical reactions with water vapor, ozone and components before falling on the underlying soil surface, and there are no annihilation processes in them. These factors are taken into account in the mathematical modeling of the diffusion process of pollution of roadside areas with solid particles.

8. The diffusion-convective mechanism of movement of pollutants was described by the standard diffusion equation with a mass source g1.

9.Methods that help in the process of determining the dispersed composition of dust are analyzed. The method of studying the level of dustiness of motor transport roads "The method of microscopic analysis of the dispersion composition of dust using a personal computer (PC)" is selected and recommended, which allows fixing dust particles from 0.1 microns to 250 microns in the air.

10.An algorithm of an improved system for monitoring the content of fine dust in the atmosphere of a residential zone is proposed.

11. As a technical measure to reduce atmospheric pollution from internal combustion engines, it is advisable to have a two-stage gas purification system consisting of a liquid neutralizer and a standard catalytic adsorber. [4,5,6]

References

[1] Delphi 2012 Worldwide emissions standards – heavy duty and off-highway vehicles (Delphi, Michigan)
[2] Heyder U, Schaphoff S, Gerten D and Lucht W 2011 Risk of severe climate change impact on the terrestrial biosphere Environ Res Lett **6** 034036
[3] Hiroyuki Y, Misawa K, Suzuki D, Tanaka K, Matsumoto J, Fujii M and Tanaka K 2011 Detailed analysis of diesel vehicle exhaust emissions: nitrogen oxides, hydrocarbons and particulate size distributions Proc Combust Inst **33** 2895–2902
[4] Zvyagintseva A V and Samofalova A S 2020 AIP Conference Proceedings **2313** 060020
[5] Zvyagintseva A V 2020 Bulletin of the Russian Academy of Sciences: Physics **84(9)** 1097–1099
[6] Zvyagintseva A V 2020 IOP Conference Series: Materials Science and Engineering **919** 62054
[7] Skripachev V O and et al 2020 Remote Sensing Applications: Society and Environment **19** 100328
[8] Skripachev V O and et al 2020 Remote Sensing **12(371)** 1–31
[9] Zvyagintseva A V 2020 International Journal of Hydrogen Energy **45(46)** 24991–25001
[10] Morozov O, Zhurba V, Neklyudov I, Mats O, Progolaieva V and Boshko V 2016 Nanoscale

Research Letters **11** 44

[11] Walsh M P 2011 Mobile source related air pollution: effects on health and the environment Encyclopedia of Environ Health **3** 803–809

[12] Huttunen S August 1979 The integrative effects of airborne pollutants on boreal forest ecosystems Proc. of the Symp. On the Effects of Airborne Pollution on Vegetation (Warsawa) pp 111–132

[13] Manion P D 1991 Tree disease concepts (New Jersey: Prentice Hall) p 402

[14] Song Zhiyao, Wu Tingting, Xu Fumin and Li Ruijie 2008 A simple formula for predicting settling velocity of sediment particles Water Science and Eng. **1(1)** 37–43

[15] Concha F and Almendra E R 1979 Settling velocities of particle systems: 1. Settling velocities of individual spherical particles International Journal of Mineral Processing **75(4)** 349–358

[16] Zigrang D J and Sylvester N D 1981 An explicit equation for particle settling velocities in solid-liquid systems AIChE J. **27(6)** 1043–1052

[17] Brown P P and Lawler D F 2003 Sphere drag and settling velocity revisited Journal of Environmental Engineering **129(3)** 222–232

[18] Camenen B 2007 Simple and general formula for the settling velocity of particles Journal of Hydraulic Engineering **133(2)** 229–238

[19] Jimenez J A and Madsen O S 2003 A simple formula to estimate settling velocity of natural sediments Journal of Waterway, Port, Coastal and Ocean Eng. **129(2)** 70–83

[20] Owen P 1960 Dust deposition from a turbulent airstreams Aerodynamic Capture of particles (New York) 6–25

IOP Conf. Series: Earth and Environmental Science **938** (2021) 012012 doi:10.1088/1755-1315/938/1/012012

The use of web technologies in the transportation of medical waste generated during the treatment of COVID-19 on the territory of the Russian Federation

V G Burlov[1], M V Mironova[2], V A Abramov[3], and S Yu Kapitsyn[1]

[1]Peter the Great St. Petersburg Polytechnic University (SPbPU), Professor at the Higher School of Technosphere Safety, Doctor of Technical Sciences, Professor, Saint Petersburg, Russia
[2]Russian State Hydrometeorological University (RSHU), student in the direction "Ecology and nature management", Saint Petersburg, Russia
[3]Russian State Hydrometeorological University (RSHU), postgraduate student in the direction "Management in technical systems", Saint Petersburg, Russia

Email: mmmw98@bk.ru

Abstract. In this study, explores the use of web technologies as a way to ensure that medical waste transportation is undertaken in a proper and safe way. Insofar as an incorrect waste sorting is regarded as a source of an epidemiological threat within further transportation, is proposed a synthesis of a waste sorting model. The model is based on a natural science approach to managerial decision-making. The model enables to identify and prove violations concerning medical waste sorting. Described four interconnected components of the model: an initial stage, when vaccination processes are not taking place; the vaccination is completed and medical waste is sorted according to the rules; the vaccination is completed, but the rules of medical waste sorting are violated; all the problems related to medical waste sorting are solved, so its transportation is possible.

1. Introduction

Since 1975 on the territory of the Russian Federation the issue of integrated waste management has been intensively considered [1]. Since the beginning of the 21st century the activities of the Russian government, e.g. in the field of waste management, have been aimed at a transition to sustainable development, the basis of which is the UN concept in this area. In 2019 medical waste underwent changes in the composition, quantity, and timing of their disposal. Coping with the increased and changed volumes of healthcare waste requires dynamic and flexible action. For example, to adjust the collection and distribution of waste, it is necessary to consider changes in their quantity and composition [2].The processing enterprises in the Russian Federation are significantly limited in capacity and cannot adapt to increasing loads from medical institutions. This fact limits the collection and transportation of waste to the places of their neutralization.

Dangerous healthcare waste generated during the COVID-19 pandemic is not limited to waste generated by healthcare facilities (e.g. vaccination points). This waste includes waste generated by people who tolerate mild forms of COVID-19, such as masks, gloves, wipes, etc. The issue of medical waste management, namely their collection and transportation, is one of the most important factors in the process of ensuring environmental safety in the field of waste management.

IOP Conf. Series: Earth and Environmental Science **938** (2021) 012012 doi:10.1088/1755-1315/938/1/012012

Thus, the subjects of the Russian Federation faced the question of ensuring the safety of collection and transportation of medical waste. This activity cannot be controlled and tracked without the use of web technologies. Web technologies contain a set of technical means that regulate and direct the activities of those who manage the transportation of waste. The use of such a technology enables a real-time monitoring of all stages of transportation. If a threat occurs at any stage, it can be neutralized.

2. Method

This article uses a method based on the synthesis of a mathematical model for the adoption of management decision. For a more detailed consideration of using web technologies, it is necessary to determine the target process. In this case, the target process is taken as an activity in the process of which medical waste is generated that has a direct or indirect interaction with infected patients, e.g. used tools after vaccination (COVID-19). Next, it is necessary to identify the problem: while vaccinating the population all waste is divided into hazard classes. Waste generated during vaccination is classified as class B. Class B waste cannot be destroyed, transported without special packaging, tamped. However, a medical staff member may place potentially hazardous waste in a container that does not correspond to its hazard class or tamp the package, which will break the tightness of transportation and turn it into a potentially dangerous object carrying the virus. In order to minimize this threat, it is necessary to work out the final stage – threat neutralization. At this stage, collection and transportation of waste is controlled via web technologies. To minimize the damage to the environment and control the entire process of generating the waste, it is crucial to build an adequate decision-making model. The model should include four interconnected technological components [3]:

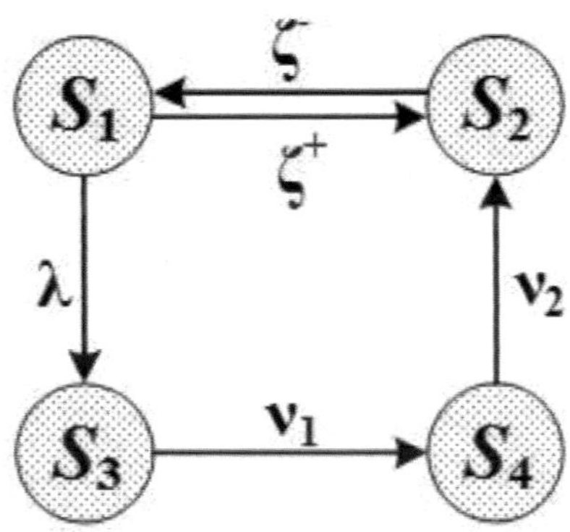

Figure 1. Management states graph

The basic states of the system have the following designations:

- S1 – the initial state of the system, vaccination is not taking place, identification and neutralization of problems is not required;
- S2 – the decision-maker (DM) has completed the target task, the waste is sorted without violations and can be transported;
- S3 – there are violations of medical waste sorting rules, a decision maker identifies the problem for its subsequent neutralization;
- S4 – neutralization of the problem, the waste is sorted correctly and can be transported.

The mathematical unit of the model [4]:

$$P = F(T_{з}, \Delta t_{пп}, \Delta t_{ип}, \Delta t_{нп}), \quad (1)$$

P – the probability of finding the control object in each of the basic states: initial, target, identification, or neutralization of problems (violations);

$T_{з}$– an average period of the target process;

$\Delta t_{\Pi\Pi}$ – an average time of problem emergence;
$\Delta t_{И\Pi}$ – an average time of problem identification;
$\Delta t_{H\Pi}$ – an average time of problem neutralization.

The basic states S_i are related to each other by the intensity of the transition: $\zeta^+ = 1/T_Э$, $\lambda = 1/\Delta t_{\Pi\Pi}$, $\nu_1 = 1/\Delta t_{И\Pi}$, $\nu_2 = 1/\Delta t_{H\Pi}$ and frequency ζ^- [5], that is characterized by disruption to the purpose. In this article, disruption is defined as the assignment of medical waste to an inappropriate class. Such a mistake is a direct consequence of the human factor, when a medical employee, due to a large number of patients or fatigue, makes a wrong decision.

The connection between the probability $P = (P_1, P_2, P_3, P_4)$ and the basic states S_i is carried out through the differential equations of Kolmogorov-Chapman provided that $P_1(t) + P_2(t) + P_3(t) + P_4(t)= 1$ [6].

It is known from [7] that at the normatively accepted levels of the maximal permissible frequency of disruptions (ζ^-) and the minimal sufficient efficiency (P_2) [8, 9], the criterion of guaranteed management of medical waste transportation allows decision makers to control the sufficiency and optimize the intensity using web technologies, making a procedural decision by rationalizing the duration of transitions by their events.

3. Results and Discussion

During the accumulation and transportation of medical waste in places of treatment and vaccination against COVID-19, the following target processes take place: prevention, detection, and neutralization. The last two processes are proposed to be carried out with the use of web technologies [10].

Due to an increase of the number of cases, the opening hours of vaccination centers were extended by 2 hours, so the current opening hours are 8:00-22:00. Moreover, a medical worker needs to prepare all the necessary equipment, so his or her working day starts averagely at 7:00. Thus, the working day of the medical personnel vaccinating the population is equivalent to 15 hours.

Table 1 shows a list of events in the network model of the target process for the prevention, identification and proving of a disrupt that occurs during the sorting of medical waste generated during vaccination of the population against COVID-19, as well as their earliest, acceptable, and possible time.

Table 1. Events of the process aimed at detection and collection of proofs of a violation of the rules of medical waste sorting

Xi	Name of the i-th event of the process of prevention, detection, and proving of violation of rules of medical waste sorting	*Tp (i) in [9], hours*	*Tn (i) in [9], hours*	*R (i) in [9], hours*
X_0	There is a medical staff member that threatens a proper medical waste sorting	0	0	0
X_1	Evidence of violation of waste sorting rules has been found.	24	24	0
X_2	Evidence of violation has been recorded via web technologies (e.g. photo and video records)	27	29	2
X_3	An authorized official has drawn up a violation report	32	36	4
X_4	Improperly sorted medical waste is removed and re-sorted for its further transportation to the places of neutralization	47	47	0

For a targeted process of preventing, detecting, and proving of a violation of rules concerning the segregation of medical waste generated during the vaccination of the population against COVID-19 the critical path through the chain of events $x_0 \rightarrow x_1 \rightarrow x_2 \rightarrow x_3 \rightarrow x_4$ is Tэ = 47 h = 1.96 days, i.e. ζ^+ =51.02 target processes per 100 days [11].

Sorting medical waste, medical staff members can create obstacles impeding to record waste sorting violations via web-based technologies.

Table 2 shows a list of events in the network model of the process of forming of a threat to a reasonable time of prevention, detection, and proving of sorting violations necessary for a further transportation of medical waste, as well as their earliest, latest, and acceptable time.

Table 2. The list of events of the network model of the process of forming of a threat to a reasonable time of prevention, detection, and proving of sorting violations necessary for a further medical waste transportation, as well as their earliest, latest and acceptable time.

Ai	Name of the i-th event of the process of formation of a threat to a reasonable time for prevention, detection, or proving of violations of rules related to medical waste sorting	*Tp (i) in [9], hours*	*Tn (i) in [9], hours*	*R (i) in [9], hours*
A_0	There is a medical staff member that threatens a reasonable time	0	0	0
A_1	Defining waste class, the employee has made a mistake and attempted to make it secret	30	30	0
A_2	Web-technology with "blind spots" has not identified the violation	52	54	2
A_3	Improperly sorted waste has gone to garbage trucks for further transportation	68	68	0
A_4	The waste has arrived at the decontamination facilities, but due to wrong information about its class contamination of all the waste occurs, which leads to further spread of viral infection.	83	83	0

For the process of forming of a threat to a reasonable time of prevention, detection, and proving of sorting violations necessary for a further transportation of medical waste, the critical path of the problem, passing through the chain of events $a_0 \rightarrow a_1 \rightarrow a_2 \rightarrow a_3 \rightarrow a_4$ is $\Delta t_{\Pi\Pi}$ = 83 h = 3.46 days, i.e. λ = 28.9 problems per 100 days [12].

Table 3 provides a list of events in the network model of the process of identifying a threat to the reasonable time of prevention, detection, and proving of violations related to collection and transportation of medical waste, as well as their earliest, latest, and permissible time.

Table 3. The list of events in the network model of the process of identifying a threat to the reasonable time of prevention, detection, and proving of violations related to collection and transportation of medical waste, as well as their earliest, latest, and permissible time.

Bi	Name of the i-th event of the process of identifying a threat to the reasonable time of prevention, detection, and proving of violations related to collection and transportation of medical waste	*Tp (i) in [9], hours*	*Tn (i) in [9], hours*	*R (i) in [9], hours*
B_0	A person responsible for decision-making concerning the management of medical waste collection and transportation with the use of web technologies has initiated collection and analysis of information.	0	0	0
B_1	Web technologies have been successfully installed and connected to the local network of medical institutions	123	126	3
B_2	Observing in real time, the person responsible for decision-making has noticed some signs of violation	49	54	5
B_3	Evidence of inaccurate medical waste sorting has been detected via web technologies (e.g. photo and video)	51	51	0
B_4	The evidence of the violation has been recorded via web technologies during the monitoring of the work of medical personnel in real time.	83	83	0
B_5	The results of identification of a threat to a reasonable time of prevention, detection, or proving of the violation have been summarized; hazardous waste has been removed from improperly sorted one, it is possible to continue waste transportation	123	123	0

For the process of identification of the threat to a reasonable time to detect, fix, and prove violations of medical waste sorting rules, the critical path through the events $b_0 \to b_1 \to b_2 \to b_3 \to b_4 \to b_5$, leaves $\Delta t_{ИП}$= 123 h = 5.12 days, i.e. $\nu_1 = 0.81$ identifications per 100 days.

Table 4 shows the events of the network model, aimed at neutralization of the threats to the reasonable time of prevention, detection, and proving of violations.

Table 4. The list of events in the network model aimed at neutralizing threats to the reasonable period of time of prevention, detection, and proving of violations

Ci	Name of the i-th event of the process of neutralization of a threat to a reasonable period of time to prevent, detect, or prove violation of medical waste sorting	*Tp (i) in [9], hours*	*Tn (i) in [9], hours*	*R (i) in [9], hours*
C_0	Web technologies have helped to establish the signs, causes, and evidence of a threat to a reasonable time of prevention, detection, and proving of violations related to the preparation of waste for transportation	0	0	0
C_1	Alternative evidence and signs of violations have been documented in accordance with the requirements of the Code Of Administrative Offences Of The Russian Federation	6	8	2
C_2	Decision makers have been obliged to prevent the transportation of incorrectly sorted waste, to correct and send correctly sorted waste to transportation.	34	34	0

For the process of neutralization of the threat to disruption of the studied processes, the critical path passing through the events $c_0 \to c_1 \to c_2$ is $\Delta t_{НП} = 47$ h = 1.42 days, i.e., $\nu_2 = 70.42$ neutralizations per 100 days.

It is assumed that an average frequency of disruptions of the target task $\beta = 0.16\zeta^{+}$ [13]. For the proportion of the time spent in the basic states of the proposed model of management of prevention, detection, and proving of violations of medical waste sorting via web technologies before transporting it to the disposal sites, the following values have been taken [14,15]: $P_1 = 0.02$, $P_2 = 0.21$, $P_3 = 0.76$, $P_4 = 0.009$ at $\zeta^+/(\zeta^+ + \zeta^-) = 0.86$. The P_2 baseline performance indicator does not significantly change if the problems connected with the installing of web technologies are taken into account [16]. This fact indicates sufficient control in the current situation, given by the intensities ζ^+ of the target process and the flow of violations λ.

4. Conclusion

This paper draws the issue of using web technologies in the transportation of medical waste generated within the process of COVID-19 treatment in the Russian Federation. Waste sorting is described in detail, since any violations of rules at this stage can be regarded as a source of an epidemiological threat along the entire process of transportation. The use of web technologies without a properly built adequate model is considered to be impossible. Only consideration of all the basic states of this model will help to minimize the risk of further spread of the virus. The calculated parameters of the intensities of the target process and the flow of violations can serve as a basis of a proper and effective use of web technologies for transportation and sorting of medical waste.

However, human activities remain the main risk factor. This factor, as well as an increasing load on the medical staff, influences the system in a negative way. Therefore, it is recommended to use web technologies not only for monitoring of all stages of waste transportation, but also for prevention and detection of violations that can be undertaken by medical staff members. Such measures are supposed to minimize the risks related to medical waste transportation.

References

[1] Babaeva A A and Grigorieva E V 2021 Household Waste Recycling Technologies *International science and technology conference "Earth science"* **720** 012005

[2] Rahayu1 P, Rohajawati S, Fairus S, Saragih H and Akbar H 2021 Challenges and Recommendation of the Information Technologies Application in Hazardous Medical Waste Management amidst Pandemic COVID-19 *Journal of Physics: Conference Series* **1844** 012029

[3] Andreev A V, Burlov V G and Grachev M I 2019 Information technologies and synthesis of the management process model in the enterprise *2019 International Science and Technology Conference "EastConf"* 8725428

[4] Burlov V, Andreev A and Gomazov F 2018 Development of a model for the management of environmental safety of the region, taking into account of the GIS capacity *MATEC Web of Conferences* **193** 02038

[5] Feller W 1957 *An introduction to probability theory and its applications*. Volume 1 second edition (New York John Wiley & Sons. Inc. London Chapman & Hall, Limited)

[6] Abramov V M, Aleshin I V, Burlov V G, Chusov A N, Istomin E P, Shilin M B and Sokolov A G 2019 Geo-information and geo-ecological support tools development for environmental economics *Proceedings of the 33rd International Business Information Management Association Conference*, IBIMA 2019: Education Excellence and Innovation Management through Vision 2020 7053–7061

[7] Burlov V G 2004 *About information war philosophy Fundamental research in technological universities*. Volume 2. Part 1. National security (Saint Petersburg, SPBGTU) pp 132–148

[8] Burlov V, Andreev A, Gomazov F and Somga-Bichoga N 2018 System integration of security maintenance processes in knowledge management *Proceedings of the European Conference on Knowledge Management* **1** 112–122

[9] Abramov V M, Aleshin I V, Burlov V G, Chusov A N, Istomin E P, Shilin M B and Sokolov A G 2019 Geo-information and geo-ecological support tools development for environmental economics *Proceedings of the 33rd International Business Information Management Association Conference*, IBIMA 2019: Education Excellence and Innovation Management through Vision 2020 7053–7061

[10] Burlov V G and Popov N N 2017 Management of the application of the space geoinformation system in the interests of ensuring the environmental safety of the region *Advances in the Astronautical Sciences* **161** 751–760
[11] Anokhin P K Systemic mechanisms of higher nervous activity *The science* p 453
[12] Burlov V G 2004 1 *Information computation system security consept in conditions of firmware distraction Fundamental research in technological universities*. Volume 2. Part 1. National security (Saint Petersburg, SPBGTU) p 226–249
[12] UNEP Waste management an essential public service in the fight to beat COVID-19 2020 *United Nation Environ. Program* 1–10
[13] Arbib M A *Brains. Machines and Mathematics* (New York: McGraw-Hill Book Co) p 494
[14] Simpson III W P and Patterson J H 1996 A multiple-tree search procedure for the resource-constrained project scheduling problem *European Journal of Operational Research* **89(3)** 525–542
[15] Lu M and Li H Resource-activity critical-path method for construction planning *Journal of Construction Engineering and Management* **129(4)** 412–420
[16] Simpson III W P and Patterson J H A multiple-tree search procedure for the resource-constrained project scheduling problem *European Journal of Operational Research* 89(3) 525–542

Geoecological and economic assessment of the municipal solid waste management system in the Kaluga region

I N Lykov[1] and E V Obolenskaya[1]

[1]Kaluga State University named after K.E. Tsiolkovsky, Kaluga, Russia

Email: linprof47@yandex.ru

Abstract. The removal of solid household waste to landfills is the most commonly used method of solid waste disposal and an important source of environmental pollution. Physical, chemical and biochemical processes occur in the body of a solid waste dump, which overlap each other and form an artificial aquifer and biogas. This article provides a brief overview of the environmental problems associated with the practice of garbage disposal and the negative impact of the landfill on the environment. The data on the features of the functioning of the solid waste landfill in Kaluga, on the dynamics of waste generation and changes in their morphological composition are presented. The issues related to the economic problems of waste disposal are considered.

1. Introduction

One of the most acute environmental problems at present is the pollution of the environment with waste. With the growth of the population, the increase in the standard of living increases the volume of household waste generation. Today, the most common method of neutralization of waste removed from a settlement is their burial at landfills of solid municipal waste (MSW). Landfills of MSW in Russia are accumulators of almost all types of waste, which is a problem for all cities and large settlements of our country without exception. Despite the ever-increasing pace of industrialization and urbanization, the disposal of solid household waste in landfills remains one of the main means of their disposal. Landfills of municipal solid waste are sources of long-term negative impact on the environment [1]. Physical, chemical and biochemical processes occur in the body of the landfill, which overlap each other and form a man-made aquifer and biogas. Filtrate and biogas are formed in the anaerobic zone of the landfill, the capacity of which reaches 10-20 m. [2]. The absence or imperfection of anti-filtration screens leads to the seepage of filtrate deep through the soil and its entry into underground water, the entry of biogas containing greenhouse gases and toxic substances into the environment. The landfill filtrate is characterized by high mineralization, a significant content of ammonium ions, heavy metals and organic compounds [3, 4, 5].

Alternative technologies for the treatment and disposal of municipal solid waste include incineration with heat recovery and purification of waste gases, composting. But both of these technologies are criticized, since they are accompanied by the release of dangerous ecotoxicants, do not eliminate pathogenic biological agents, and are unable to immobilize heavy metals [6, 7]. Therefore, landfills remain the most economical waste disposal method worldwide and the most common waste disposal method in Russia. But this leads to an irretrievable loss of up to 90% of useful products that are in real demand in the market [8]. For example, in the United States, the gross lost commodity value of recyclable materials in 2015 amounted to $ 1.4 billion [9, 10]. And one ton of waste is up to 80 kg of polymers, about 60 kg of glass, 25 kg of ferrous metals and 170 kg of paper [11].

The MSW landfill should be a profitable enterprise. However, the primitive system of operation, which turns it into a banal garbage dump, intensive environmental impact reduces the economic indicators of the landfill [12]. This, in turn, leads to financial costs in the form of fees for environmental pollution, to higher tariffs for the population and social tension. In the public consciousness, the MSW landfill is perceived as an extremely aggressive object [13, 14].

Landfills of MSW in Russia are accumulators of almost all types of waste, although many production and consumption waste could be sent for recycling. By the beginning of the 1980s, the annual output of MSW in Russia was estimated at 40 million tons, currently they are formed more than 60 million tons per year. The specific annual output of MSW per inhabitant of a modern city is 250-700 kg. In developed countries, this value increases by 4-5% annually, which is three times higher than the population growth. For example, every family in the EU countries on average throws out about 500 kg of household garbage per year. In general, the European Union produces up to 3 billion tons of solid waste annually [15].

More than 90% of all waste is disposed of in landfills and forms a significant environmental load due to the withdrawal of land resources, emissions into the atmosphere and water bodies at waste disposal sites [16]. Therefore, the purpose of our study was to study the environmental and economic problems associated with the formation, collection and disposal of MSW as a secondary raw material using the example of the Kaluga region. The scientific novelty of the research is associated with new environmental and economic information, which will help in the further development of a resource strategy for MSW management.

2. Objects and methods of research

The main method of this study is the process of processing and analyzing environmental and economic information related to the accumulation and disposal of MSW. The object of our research was the solid municipal waste landfill of the city of Kaluga. The total area of the landfill is 15 hectares, the roof is 6.77 hectares, the slopes are 14.2 hectares. The volume of buried waste is 8155172.2 m^3, or 1190012.6 tons. The landfill has been in operation since 1972. In the section, the landfill is a block of different-age layers of solid waste, interspersed with soil. The height of the polygon body today reaches 25-30 m.

3. Research results and discussion

In 2016, more than 500 thousand tons of solid municipal waste were generated on the territory of the Kaluga Region, including more than 325 thousand tons from the population. As a result, the indicator of the number of MSW of the population to the total population was 324 kg per 1 inhabitant [17]. This is the average indicator in the Central Federal District (Fig. 1).

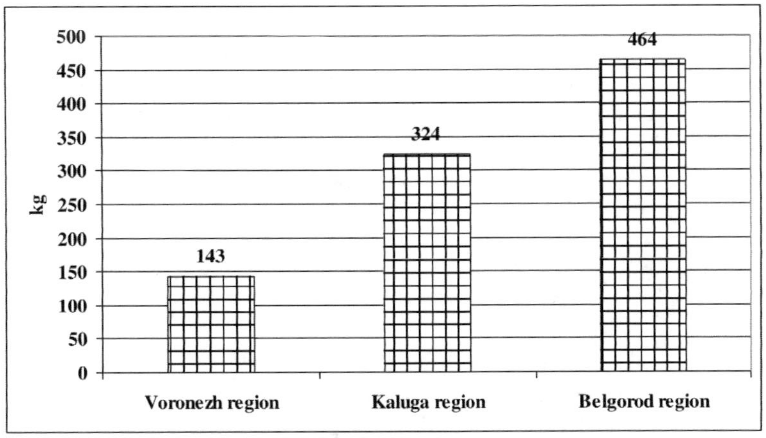

Figure 1. The number of MSW per one inhabitant of the Kaluga region

The bulk of solid municipal waste comes from the north-eastern part of the Kaluga region, which is characterized by the greatest development of industry and agriculture, the concentration of the main part of the population. Waste that is characterized by a high level of liquidity in the current market conditions is mainly collected and disposed of as secondary raw materials. These include glass, waste of ferrous and non-ferrous metals, plastic, paper, cardboard, which makes up 39.5% of household waste. From the total annual amount of production and consumption waste by enterprises, from 81.1% to 96% were used (Table 1). Also, 2054.12 tons of waste were neutralized, which is 0.22%. The main share of waste to be neutralized is mercury lamps, spent fluorescent mercury-containing tubes [12].

Table 1. The dynamics of the accumulation and use of waste in the Kaluga region (thousand tons)

Waste	Observation time, years					
	2010	2012	2014	2016	2018	2020
Production and consumption	1846.3	4604	5130	4924	4850	4760
Of them used	1739.4	4231	4849	3992	4100	4300
Percentage of use	83.3	92	94.5	81.1	84.5	90,3
Households and municipal	1817	4546	5074	5125	5340	5780
Of them used	59.96	159.1	187.7	189.6	191.3	195.7
Percentage of use	3.3	3.5	3.7	3.69	3.57	3.38

Also, 2054.12 tons of waste were neutralized, which is 0.22%. The main share of waste to be neutralized is mercury lamps, spent fluorescent mercury-containing tubes [12].

The use of recyclable materials allows enterprises to seriously reduce costs. For example, the savings from the use of recyclable materials in the production of packaging is on average 20% of the cost price. And the cost reduction when using cullet for glass production can be up to 50%.

A characteristic feature of the waste generation structure in the Kaluga region is a low proportion of hazardous waste. Based on the data of the state statistical reporting for 2019, the share of waste generation of I class of hazard for the environment of the total volume of waste generated is 0.04%, II class - 0.017%, III class - 0.094%, IV class - 43.91%, Class V - 55.93% [17].

Solid waste management is the fastest growing segment of the environmental sector of the economy. This is especially true in the regions where the best high-tech and innovative organizations using advanced production technologies place their production facilities. Such regions include the Kaluga Region, the administrative and industrial regional center of which is the city of Kaluga with a population of about 350 thousand. The city hosts a large number of industrial, small and medium-sized enterprises, institutions, shops, public catering enterprises, parks and other objects where waste is generated. On average, more than 1.0 million m3 of waste is produced annually in the city. Most of the waste is household, institutional and commercial waste (Table 2).

The main place of waste accumulation in the city of Kaluga is the MSW landfill, which has existed since 1972. The total area of the landfill is 17.2 hectares. Due to the increase in the volume of waste entering the landfill, the filling of the areas is intensive. Geomorphologically, the site is confined to a gently undulating, dissected water-glacial plain. The absolute marks of the earth's surface range from 172.0 m to 189.0 m. Quaternary and carboniferous deposits take part in the geological structure of the site to the explored depth.

Table 2. Average annual standard for the accumulation of municipal solid waste in Kaluga

No. p/p	Object name	Unit of measurement	Average annual accumulation rate (kg)
1	Household waste	per inhabitant	from 361.0 to 445.1
2	Construction waste	per 1 m^2	168.0
3	Institutional waste (school and preschool, hospitals, universities, libraries)	per inhabitant	from 37.7 to 492.1
4	Commercial waste (shops, offices, restaurants, hotels, markets, warehouses, ateliers, hairdressers, etc.)	1 place	from 20.7 to 240.0
5	Railway stations	per 1 m^2	136.5
6	Gardening associations	for 1 site	576.0
7	Industrial enterprises	for 1 employee	180.5

The main place of waste accumulation in the city of Kaluga is the MSW landfill, which has existed since 1972. The total area of the landfill is 17.2 hectares. Due to the increase in the volume of waste entering the landfill, the filling of the areas is intensive. Geomorphologically, the site is confined to a gently undulating, dissected water-glacial plain. The absolute marks of the earth's surface range from 172.0 m to 189.0 m. Quaternary and carboniferous deposits take part in the geological structure of the site to the explored depth.

MSW landfill refers to unorganized sources of atmospheric pollution. Complex organic substances, which mainly consist of solid household waste, decompose over time in the thickness of the landfill, but the processes are very slow. Initially, there are aerobic processes under the influence of microflora, which uses oxygen for the oxidation of organic substances. The products of aerobic decomposition of MSW are carbon dioxide, nitrates and water [14, 15]. With the depletion of oxygen reserves in the lower horizons of the landfill, anaerobic processes of splitting complex organic substances, mainly hydrocarbons and complex organic acids, with the formation of organic acids, hydrogen, ammonia, carbon dioxide, methane and water, begin to prevail). 1.5 cubic meters of gases are released from one cubic meter of solid household waste.

The main share of waste buried at the landfill is household and municipal waste (63.7 % of the total volume of buried waste), food production waste (10.8 % of the total volume of buried waste), paper and cardboard waste (10.3% of the total volume of buried waste). These wastes end up at the landfill, polluting the environment (Table 3).

IOP Conf. Series: Earth and Environmental Science **938** (2021) 012013 doi:10.1088/1755-1315/938/1/012013

Table 3. Average annual data of the morphological composition of solid waste in Kaluga

Composition of solid municipal waste	Summer, %	Autumn, %	Winter, %	Spring, %	Average annual value, %
Paper, cardboard	29.3	21.8	17.6	27.0	23.9
Food waste	29.7	34.6	39.5	35.7	34.9
Tree, leaves	3.5	7.4	1.8	3.1	4.0
Textile	4.2	5.3	4.7	4.4	4.6
Leather, rubber	1.7	2.8	3.4	2.4	2.6
Polymers	5.0	4.9	3.1	3.4	4.1
Bones	0.9	2.1	1.3	1.8	1.5
Metal black	5.6	3.5	3.9	4.1	4.3
Non-ferrous metal	0.4	0.3	0.7	0.6	0.5
Glass	5.0	5.2	10.2	5.4	6.7
Stones, ceramics	1.5	2.8	1.1	1.9	1.8
Crushed stone less than 16 mm	13.2	8.3	12.7	10.2	11.1
TOTAL	100.0	100.0	100.0	100.0	100.0

Physical, chemical and biochemical processes occur in the body of the landfill, which overlap each other and form a man-made aquifer [20], the basis of the balance of which is infiltration waters fed by precipitation (about 60% of the total precipitation). Modern waste disposal technologies, the use of materials or energy from waste by thermal, chemical or biological methods, require either sufficient investment or a significant reduction of waste on the part of the population. The recovery of one ton of MSW using environmentally acceptable technologies requires at least 70-100 US dollars.

The bulk of food waste, paper and paper products is mainly represented by C^3-plants. About 7% of the waste is made up of polymeric materials and rubber products, for the complete mineralization of which a long period of physicochemical and biological effects is required. About 72% of MSW can be biodegradable with the dominant participation of the microbial community. The end result of biodegradation of waste by a mixed microbial association is the formation of biogas, which is a mixture of 65-75% methane and 20-35% carbon dioxide, which are included in global climatic processes. Biogas also contains minor amounts of hydrogen sulfide, nitrogen, hydrogen and argon.

As a result of biodegradation, a sufficient amount of water is also formed, which in the form of landfill condensate enters surface and groundwater, contributing to the migration of pollutants.

The carbon content in the waste stored at the MSW landfill (Kaluga), which can be used by microorganisms as a substrate for biogas production, is estimated at about 14% of the annually received organic waste. Consequently, when storing organic products about 21.87×103 t / year, the amount of carbon used by microorganisms for biogas production will be 3.06×103 t C / year, which is equivalent to the formation of biogas in an amount of about 5.6×106 m^3 / year.

We revealed an extremely uneven release of methane from the landfill body: from 0 to 3.9 liters per hour from a square meter of the surface. This reflects the local condition of the clay layer overlying the waste layers. The main flow of methane into the atmosphere occurs on the slopes of the landfill and is practically not observed on its roof (Table 4).

Biogas generated in the bulk of the stored waste spreads mainly in the horizontal direction, and the emission situation depends mainly on the state of the upper layer of waste and the degree of their isolation from atmospheric air.

Table 4. Change in the composition of biogas depending on the location and depth of the well

Well placement, depth	Amount, %			
	CH_4	CO_2	N_2	O_2
on the slope of the landfill, 15 m	44.1	23.8	26.0	6.2
on the slope of the landfill, 10 m	59.1	34.3	6.0	1.0
on the slope of the landfill, 5 m	63.4	22.8	11.1	2.6
on the roof of the landfill, 10 m	20.8	14.5	68.8	13.3
on the roof of the landfill, 15 m	10.2	5.2	72.2	12.3

Large MSW landfills will require more investments to ensure environmental safety than small facilities. However, the specific investment (per ton of landfilled waste or per capita) will decrease with increasing site size. There are financial and other benefits for MSW landfills with a long service life (ten or more years). Large regional MSW landfills serving two or more cities can be cost-effective as the costs of transporting waste are reduced.

Optimization of the processes of waste collection, sorting and disposal of their residues will speed up the processes of biodegradation and increase the economic efficiency of landfills. In addition, the technical and economic performance of MSW landfills is greatly influenced by the processes of generating biogas and its main component – methane [18].

References

[1] Vaverková M D, Adamcová Dana, Zloch J, Radziemska M, Berg A B, Voběrková S and Maxianová 2018 A Impact of Municipal Solid Waste Landfill on Environment – a Case Study *J. Ecol. Eng.* **19(4)** 55–68 DOI: 10.12911/22998993/89664

[2] Kuo J and Dow J 2017 Biogas production from anaerobic digestion of food waste and relevant air quality im plications *Journal of the air & Waste management association* **67(9)** 1000–1011 DOI: 10.1080/10962247.2017.1316326

[3] Peng Yao 2017 Perspectives on technology for landfill leachate treatment *Arabian Journal of Chemistry* **10(2)** 2567–2574

[4] Vaverková M D, Elbl J, Koda E, Adamcová D, Bilgin A, Lukas V, Podlasek A, Kintl A, Wdowska M, Brtnický M and Zloch J 2020 Chemical composition and hazardous effects of leachate from the active municipal solid waste landfill surrounded by farmlands *Sustainability* **12** 4531 DOI: 10.3390/su12114531

[5] Naveena B P, Mahapatrab D M, Sitharama T G, Sivapullaiaha P V and Ramachandra T V 2016 Physico-chemical and biological characterization of urban municipallandfill leachate *Environmental Pollution* DOI: 10.1016/j.envpol.2016.09.002

[6] Lopes E J, Okamura L A and Yamamoto C I 2015 Formation of dioxins and furans during municipal solid waste gasification *Environmental engineering. Braz. J Chem. Eng.* **32(1)** DOI: 10.1590/0104-6632.20150321s00003163

[7] Smith St R A 2009 Critical review of the bioavailability and impacts of heavy metals in municipal solid waste composts compared to sewage sludge *Environ Int.* **35(1)** 142–156 DOI: 10.1016/j.envint.2008.06.009

[8] Abdel-Shafy H I and Mansou M S M 2018 Solid waste issue: Sources, composition, disposal, recycling, and valorization *Egyptian Journal of Petroleum* **27(4)** 1275–1290 DOI: doi.org/10.1016/j.ejpe.2018.07.003

[9] Powell J T, Pons J C and Chertow M 2016 Waste Informatics: Establishing Characteristics of Contemporary U.S. Landfill Quantities and Practices *Environ. Sci. Technol.* **50(20)** 10877–10884 DOI: 10.1021/acs.est.6b02848

[10] Powell J and Chertow M 2019 Quantity, Components, and Value of Waste Materials Landfilled in the United States *Journal of Industrial Ecology* **23(2)** 466–479 DOI: 10.1111/jiec.12752

[11] Waste in Russia: garbage or a valuable resource? Availble at: tpprf.ru/download.php?

[12] Ozcan H K, Guvenc S Y, Guvenc L and Demir G 2016 Municipal Solid Waste Characterization according to Different Income Levels: A Case Study *Sustainability* **8** 1044 Available at: www.mdpi.com/journal/sustainability
[13] Njoku P O, Edokpayi J N and Odiyo J O 2019 Health and environmental risks of residents living close to a landfill: a case study of thohoyandou landfill, Limpopo Province, South Africa *Int J Environ Res Public Health* **16(12)** 2125 DOI: 10.3390/ijerph16122125
[14] Ferronato N and Torretta V 2019 Waste mismanagement in developing countries: a review of global issues *Int J Environ Res Public Health* **16(6)** 1060 DOI: 10.3390/ijerph16061060
[15] On the progress of work to identify and compel the liquidation of places of unauthorized placement of solid household waste Report of the Federal Service for Supervision of Environmental Management Available at: http://rpn.gov.ru/node/14
[16] Valavanidis A and Vlachogianni Th 2015 Municipal Solid Waste and Environmental Pollution. Trends of Municipal Waste Management in European Countries and in Greece. Project: Plastic waste, microplastic pollution in the aquatic environment Available at: www.researchgate.net/publication/273888008
[17] Report on the state of natural resources and environmental protection in the Kaluga Region in 2019 (Izhevsk: OOO "Print") 2020, p 300
[18] Lykov I N, Zyakun A M, Obolenskaya E V and Golofteeva A Since 2016 Waste disposal at the MSW landfill in Kaluga and the intensity of biogas formation *Ecology of urbanized territories* **4** 44–49
[19] Rastogi M, Nandal M and Khosla B 2020 Microbes as vital additives for solid waste composting *Heliyon* **6(2)** e03343 DOI: 10.1016/j.heliyon.2020.e03343
[20] Iravanian A and O Ravari Sh 2020 Types of Contamination in Landfills and Effects on The Environment: A Review Study *IOP Conf. Series: Earth and Environmental Science* **614** 012083 DOI:10.1088/1755-1315/614/1/0120831

The Role of Secondary Resource Market in the Development of the Extended Producer Responsibility Mechanism in Russia

M A Liubarskaia[1] and N A Putinceva[2]

[1]Department of State and Territorial Planning, Saint-Petersburg State University of Economics, 21 Sadovaya Street, 191023 Saint-Petersburg, Russia
[2]High School of Engineering and Economics, Peter the Great Saint-Petersburg Polytechnic University, 29 Polytechnic Street, 195251, Saint-Petersburg, Russia

E-mail: lioubarskaya@mail.ru

Abstract. The paper is aimed at the evaluation of current state of secondary resource market in Russia and the determination of factors of their influence on the extended producer responsibility (EPR). Research objectives include scientific justification of directions for the improvement of solid waste management handling process in the Russian regions in order to provide the support for manufactories and importers performing the extended producer responsibility activities. GAP-analysis was used as a method of obtaining the systematic results on the basis of the information received about the current state of the Russian secondary resource market according to various criteria, including the product quality, processing methods, communication between producers and consumers, and state regulations. Research is executed on the grounds of collaborative use of positivistic and phenomenological approaches, where the findings are based on the comparison of experts' opinions on the factors influencing the development of EPR mechanism, and statistics of the financial performance of main players of the Russian secondary resource market. As a scientific novelty, it is proposed to use the development of secondary resource market for enhancing of EPR mechanism in Russia. The study reveals that successful functioning of the extended producer responsibility mechanism depends not only on business representatives involved in the production of goods and packaging, but also on consumers and on the state authorities.

1. Introduction

Extended producer responsibility (EPR) is considered worldwide as a financial mechanism for the development of waste management systems. This mechanism plays an important role in the transition to the principles of a circular economy and environmentally responsible production. Since the manufacturers saturate the markets with goods and packaging forming the waste flows in the process of consumption, it depends on them how suitable the waste will be for subsequent processing.

According to Maitre-Ekern E [1], sustainable circular economy supports the second-hand trading (reuse) and involvement of secondary materials in the production cycle. To achieve the goals of increasing recycling rates, countries around the world came to the solution of enhancing participation of producers and importers of certain categories of goods in the financial and organizational aspects of waste management systems.

In Russia, as well as in the European Union and other countries of the world, manufacturers and importers who bring goods and packaging to the national markets are recognized as "polluters" of the environment, and they are responsible for paying the utilization fee. Such amendments were made to

IOP Conf. Series: Earth and Environmental Science **938** (2021) 012014 doi:10.1088/1755-1315/938/1/012014

the Federal Law 89 "On Waste of Production and Consumption" at the end of 2014, and the EPR mechanism was introduced into the legislation of the Russian Federation.

The proper performance of participants of EPR mechanism should contribute to the implementation of the national project "Ecology" aimed at improving the environmental situation in the Russian regions by increasing the efficiency of waste management systems.

According to the Ministry of Natural Resources and Environment of the Russian Federation, in 2019, 15,000 enterprises out of 160,000 operating in Russia reported on the utilization of goods and packaging after they lost their consumer properties, and as part of the payment of the utilization fee, the budget of the Russian Federation collected 3.7 billion rubles. At the same time, the population of Russia pays 193 billion rubles for waste disposal annually, which is incomparably more. As a result, the role of EPR in financing the waste management system is currently minimal. Many enterprises themselves cannot organize the utilization of goods and packaging after they are out of use, and finding consumers interested in these secondary resources is a challenge due to the underdevelopment of the market.

2. Materials and Methods

Paper presents the combination of positivistic and phenomenological approaches, where the findings are based on the comparison of experts' opinions on the factors influencing the development of EPR mechanism, and GAP-analysis of the Russian secondary resource market.

As a scientific novelty, it is proposed to use the development of secondary resource market for enhancing of EPR mechanism in Russia.

The market for secondary material resources not only plays a vital role in the sustainable development of the spheres of production and consumption, as elements of a modern economy, during its transition to a cyclical model of functioning, but also makes a significant contribution to reducing the negative impact of the waste management sector on the environment, while positively impacting the implementation of extended producer responsibility.

As a relatively stable, modern world market for secondary material resources contributes to the sustainable development of the secondary processing industry and attracting the interest of investors to it. The GAP-analysis of the Russian market of secondary material resources was conducted. The main purpose of GAP-analysis is to identify those market opportunities that can become effective market advantages. This method is a tool that can clearly describe the existing situation and the one that needs to be achieved.

GAP-analysis is a comprehensive analytical study that explorers the inconsistencies, gaps between the current state of a company, industry or market and the desired one. This analysis also allows to identify problem areas ("bottleneck") that impede development, and assess the degree of readiness to make the transition from the current state to the desired one.

In the course of the research, the GAP-analysis was carried out on the basis of the information received about the current state of the market according to various criteria. This process included the following steps:

- evaluation of current state of secondary resources market in Russia and assessment of the general gap;
- division of general gap into components depending on the functional characteristics of the market, identifying characteristics and grouping them the way necessary to conduct a review;
- identification of market opportunities to bridge the gap's components;
- development of a set of measures to achieve the set goals.

The main groups of "gaps" of the Russian secondary resource market include:

- product quality;
- processing methods;
- communication between producers and consumers;
- state regulation.

IOP Conf. Series: Earth and Environmental Science **938** (2021) 012014 doi:10.1088/1755-1315/938/1/012014

3. Results

The hierarchy of waste management methods in Russia is officially focused on the maximum use of recycled materials in production processes. But at the moment, the contribution of the waste management industry to the gross domestic product of the Russian Federation is very low (about 3%). Currently, a large amount of valuable secondary resources remain in the flow of municipal solid waste (MSW) and going to landfills.

In table 1, the description of the current status of the Russian market of secondary materials is presented, which allowed the authors to characterize the gap between the actual and the desired state of this market.

Table 1. GAP-analysis of the Russian market of secondary material resources.

Gap	Current state	Desired state
Product quality	The existing solid waste collection system is organized in the way that in most cases does not allow obtaining a secondary material of the required quality. Unitary collection of MSW components prevails, which negatively affects the possibility of further utilization of each type of material.	Introduction of separate collection for MSW and industrial waste allowing to obtain a high-quality secondary material flows.
Processing methods	Processing of MSW is organized not in all regions of Russia. Methods of waste disposal at landfills prevail, while in most cases even the energy potential of the generated landfill gas is not used.	Development of innovative technologies for the processing of certain materials, first of all, commercially demanded. These include various types of plastics, glass, home appliance components and electronics.
Communication between producers and consumers	The relationship between the producers and consumers of secondary materials has not been established.	Establishment of sustainable communications between enterprises generating flows of secondary materials and potential consumers, for example, in the form of a secondary resource exchange.
State regulation	State regulation is carried out by collecting fees from waste producers for negative impact on the environment. Amendments to the Federal Law No. 89-FZ " On Waste of Production and Consumption" now allow manufacturers to reduce the total volume of waste, which is charged for negative environmental impact, by the amount of waste transferred for disposal.	Stimulation of manufactories and importers to choose the environmentally sound materials for goods and packaging to form the desirable secondary recourse flow. Introduction of ban on landfilling of certain types of waste and materials.

The implementation of the measures presented in table 1 will create new drivers for the development of the market of secondary material resources in the regions of Russia and will provide:

- cut of waste flow going to landfills, and, accordingly, reduction of negative impact on the environment and the health of the population;
- formation of regional capacities for the processing of secondary material resources;

- achievement of economic and social stability in the regions where new jobs will be created in the field of waste processing and their transfer to the category of secondary material resources;
- increasing the resource availability of industries located in the regions, where secondary materials will be more actively involved in the production turnover.

4. Discussion

As the analysis of numerous researches shows, positive processes in the economy and society, such as economic growth, increased consumption of goods and the development of trade, have their downside. Over time, among other market failures, they lead to the growth of environmental pollution. Bian Y, Song K and Bai J [2] consider the economic progress as a main factor of increasing environmental problems.

The challenge for modern academic economists in various parts of the world is to use market mechanisms to solve environmental problems. One of such mechanisms is the extended producer responsibility, implemented in many countries [3, 4]. By imposing the financial responsibility for the utilization of goods and packaging after they have been disused, state authorities are trying to involve the private enterprises in the process of secondary resource circulation.

Tian X, Xiao H, Liu Y and Ding W [5] stated that the paradigm shift from the linear to circular economy improves resource productivity, benefits businesses and society, creates employment opportunities and ensures environmental sustainability and resource circulation. Within the circular economy paradigm waste could be used as a source of secondary material or energy [6]. To be able to perform the EPR duties more efficient, producers need to see the demand for secondary materials in the relevant markets, otherwise they will just formally collect goods and packaging after they have been disused, but not really utilize them as resources.

As international experience showing, the cooperation between the manufactories and solid waste management companies helps to create effective collaborations that can organize the secondary resource supply chains and develop relevant markets [7, 8]. In the Russian solid waste management system recently an important role moved to the Regional operators, introduced by the amendments to the Federal Law No. 89-FZ "On Waste of Production and Consumption" in 2014. This tendency is clearly seen from the analysis of the financial data of the players of secondary resource market, presented at table 2.

Table 2. TOP-10 organizations by revenue performing waste collection and processing of secondary materials (Russian Federation, 2019-2020) [a]

	Revenue, 2020 (mln. RUB)	Revenue, 2020 (mln. RUB)	Assets, 2020 (mln. RUB)	Assets, 2020 (mln. RUB)	Region
JSC "Russian Copper Company"	90 653	58 186	187 521	134 229	Sverdlovsk region
LLC "Vtorresurs-Pererabotka"	51 101	32 390	15 378	5 892	Kemerovo region
JSC "Pskovvtormet"	18 309	18 309	7 066	7 066	Pskov region
LLC "OMK-EcoMetall"	17 055	17 559	2 193	2 116	Moscow
LLC "Neva-Promtorg"	14 184	n/a	6 006	n/a	Saint-Petersburg
LLC "Mosvtormet"	11 204	10 583	2 953	1 904	Moscow
LLC "Charter"	11 126	9 204	5 291	5 314	Moscow
LLC "Industrial Operator KTZ"	11 110	n/a	7 472	n/a	Chuvashia
LLC "Vtormetresurs"	10 227	n/a	0, 522	n/a	Moscow region
LLC "EcoLife"	8 982	7 615	12 508	8 735	Moscow

[a] Based on the information from https://www.testfirm.ru/rating/38/

The revenue and expenses of the majority of Russian companies performing waste collection and processing of secondary materials are growing, which indicates that this market segment is promising.

The process of enhancing of secondary resource circulation in many countries also involves introduction of ban on landfilling of certain types of waste and materials [9]. In the Russian Federation the main steps towards banning the valuable recourse landfilling started in 2018, as presented at the figure 1.

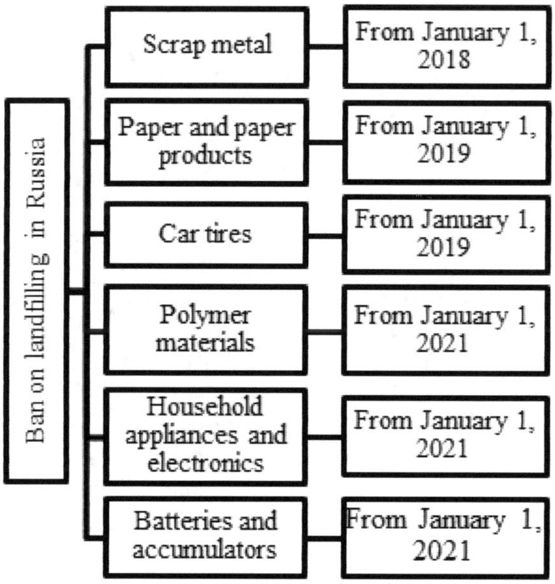

Figure 1. The phased introduction in the Russian Federation of a ban on the landfilling of certain types of secondary material resources.

Uncertainty in future prices affects all markets, but recycled materials appear to have relatively high price volatility compared to primary resources [10, 11]. The lack of a stable base for predicting future prices of recycled materials affects both the short and long term resource allocation [12]. In the short term, suppliers of secondary material resources and consumers face great difficulties in planning their market activities [13]. Without reliable information on future prices for certain types of secondary resources, decisions on their production and stocks cannot be made rationally [14, 15, 16]. As a result, manufacturers prioritize primary resources over secondary resources, even when quality and other characteristics are comparable.

For a more active development of the Russian market of secondary material resources, it is necessary:

1) to create an infrastructure for the efficient separate collection of secondary resources by type;

2) to increase the level of environmental fee rates for those who do not want to deal with the utilization of secondary raw materials (since at present the level of collection of environmental fees is much lower than planned, which limits the possibility of financing measures to increase the recycling of waste);

3) to form a stable demand for secondary material resources by stimulating producers to involve them in their technological cycles;

4) to ensure the availability of reliable models for forecasting the price and volume characteristics of the secondary materials market;

5) to develop and approve state programs that stimulate investment in the formation of a cyclical economy, including through the development of a market for secondary material resources;

6) to establish the relationship between the environmental and economic characteristics of companies' activities, including by taking into account ESG indicators when issuing bank loans and quotation of shares on the stock exchange.

5. Conclusions

As shown by the study, the prospects for the development of extended producer responsibility are determined by a number of factors in the Russian market of secondary materials, including regulatory, economic, social, environmental, and technological.

On the part of regulatory and legal factors, the strengths are the establishment of the priority in the national project "Ecology" for the allocation of secondary resources from the waste flows, as well as the establishment of 100% solid waste recycling rate as one of the national development goals in the Decree of the President of the Russian Federation. Among the weaknesses, it should be noted the slow process of approval of the Federal law "On secondary material resources" and the lack of sanctions against enterprises for failure to comply with waste recycling and utilization standards.

Positive economic factors include the presence in the national project "Ecology" of allocated state funding for the implementation of measures for the separate MSW collection and segregation of secondary materials flows, and the formation of a pool of investors willing to put money in the secondary resource utilization projects. At the same time, the development of the Russian secondary resource market is hampered by the unstable prices for reusable materials and the price imbalance in Russia's foreign trade activities in the world market of secondary resources (Russia imports more expensive secondary resources than it exports).

Among social factors, the growth of public understanding of the impact of solid waste management sphere and secondary resource market on the state of the environment and public health, as well as the interest of entrepreneurs in developing their own business in the field of collection and utilization of secondary materials, especially among those who are experiencing difficulties in other sectors of the economy in connection with the COVID-19 pandemic. The weaknesses include the lack of historical experience of separate waste collection and a low level of awareness of the rules for the collection and utilization of secondary materials among the population and organizations.

The promotion of the concept of corporate environmental responsibility in business circles makes it possible to build more optimistic forecasts for the development of the Russian market of secondary material resources, taking into account the negative consequences of the functioning of existing infrastructure facilities for waste management on the environment and public health. With the emergence of a sustainable market, the principles of infrastructure functioning should change, shifting priorities from waste disposal to utilization as secondary resources.

The emergence of Russian and foreign technologies for the processing of secondary materials on the market, allowing to obtain an ecological and economic effect, and the formation of databases of the best available technologies (BAT) create a technological basis for the development of the market for secondary materials. However, along with this, there is the use of technologies with an insufficiently high environmental and economic effect at existing facilities in the waste management sector and low rates of BAT implementation in Russian regions.

The regulatory requirements imposed on manufacturers and importers by the legislation of various countries determine the amount of their financial liability for usage of certain materials and production technologies and organizing the disposal of goods after the loss of consumer properties. However, the successful functioning of the extended producer responsibility mechanism depends not only on business representatives involved in the production of goods, but also on consumers and on the state. Such position is supported by the results of the research of Pazoki M and Zaccour G [17], where they stated that available waste management infrastructure and consumers' environmental awareness influence the return rate of used products.

It is consumers who, by their choice of goods, "vote" for the predominance of certain materials in the market, and the government sets the "rules of the game" in the commodity markets. In order for manufacturers to be able to recycle goods more successfully after the loss of consumer properties, it is necessary to develop markets for secondary resources. Only in this case, the collection of goods after they have lost their consumer properties will not be formal, but will have the goal of promoting the collected waste to the secondary market as a valuable raw material for other producers.

References

[1] Maitre-Ekern E 2021 *J. Clean. Prod.* **286** DOI: 10.1016/j.jclepro.2020.125454
[2] Bian Y, Song K and Bai J 2019 *J. Clean. Prod.* **228** 376–387
[3] Gu Y, Wu Y, Xu M, Wang H and Zuo T 2017 *J. Clean. Prod.* **164** 347–356
[4] Xu J, Ye M, Lu W, Bao Z and Webster C 2021 *J. Clean. Prod.* **282** DOI: 10.1016/j.jclepro.2020.124540
[5] Tian X, Xiao H, Liu Y and Ding W 2021 *Waste Manag.* **126** 78–88
[6] Hyun M, Kim Y and Eom J 2020 *Energy Pol.* **147** 111–119
[7] Milios L, Christensen L H, McKinnon D, Christensen C, Rasch K and Eriksen M H 2018 *Waste Manag.* **76** 180–189
[8] Schreck M and Wagner J 2017 *Waste Manag.* **67** 354–359
[9] Rene E R, Sethurajan M, Ponnusamy V K, Kumar G, Dung T N B, Brindhadevi K and Pugazhendhi A 2021 *J. Haz. Mat.* **416** DOI: 10.1016/j.jhazmat.2021.125664
[10] Liao M I, Shin X H and Ma H W 2019 *J. Clean. Prod.* **239** DOI: 10.1016/j.jclepro.2019.118142
[11] Kumamaru H and Takeuchi K 2021 *Waste Manag.* **126** 360–366
[12] Prakash S, Wijayasundara M, Pathirana P N and Law K 2021 *Res. Cons. Rec.* **168** DOI: 10.1016/j.resconrec.2020.105312
[13] Tian X, Gao W, Liu Y and Xu M 2020 *Res. Cons. Rec.* **161** DOI: 10.1016/j.resconrec.2020.104958
[14] Lonca J, Lesage P, Majeau-Bettez G, Bernard S and Margni M 2020 *Res. Cons. Rec.* **162** DOI: 10.1016/j.resconrec.2020.105013
[15] Panda S and Akcil A 2021 *Waste Manag.* **123** 48–51
[16] Akkalatham W and Taghipour A 2021 *Env. Chal.* **4** DOI: 10.1016/j.envc.2021.100112
[17] Pazoki M and Zaccour G 2021 *J. Clean. Prod.* **292** DOI: 10.1016/j.jclepro.2021.125966

Mining wastes of the Russian North-West: microbiological prerequisites of reclamation

E Abakumov[1,2], A Kimeklis[1,2], G Gladkov[1,2], E Andronov[1,2], A Zverev[1,2]

[1]Saint-Petersburg State University, 199178, 16 line V.O., Saint-Petersburg, Russia
[2]All Russia Institute for Agricultural Microbiology, 196608, Podbelskogo 3, Pushkin-Saint-Petersburg, Russia

E-mail: e_abakumov@spbu.ru

Abstract. Russian North-West is the region with extremely high diversity of superficial mining wastes, deposited in form heaps and hills. These substrata are potentially dangerous for environments and. therefore, should be reclaimed. One of the most effective way of utilization of mineral wastes is to incorporate them to natural processes of revegetation and to initialize the priming of initial soil formation process. Thus, we have investigated microbial processes in very initial superficial ecosystems of numerous heaps of pre quaternary materials of former abandoned and recently explored quarries. The research question of this study: what groups of microorganisms we do have on the surface of mining mineral wastes and are we able to qualify their role in further recycling of mineral waste to natural processes, including pedogenesis and ecogenesis of terrestrial ecosystems. The great variation of taxonomy composition of microbiome composition was revealed. At the same time, some phylums of microorganisms were dominant in all revegetated heaps investigated - Proteobacteria, Actinobacteria, Acidobacteria, Bacteroidetes, Chloroflexia, Verrucomicrobia. They could be considered as a core components of very initial soil microbiome. Primary data have showed the increasing of microbial biodiversity with timing rate of waste exposure on surface.

1. Introduction

More than 80 billion tons of mining and industrial wastes have been accumulated in Russia, and according to the most modest estimates, this volume increases by 2 billion tons annually. The solution to the problem of recycling these rocks is either the recycling of substances contained in dumps and similar formations, or the return of the corresponding material components to the biological cycle [1-4]. The latter process is closely related to the reclamation and rehabilitation of man-made dumps in order to reduce the risks of negative environmental effects. Recultivation and reclamation of disturbed lands, mountain dumps and man-made wastes as a scientific direction is relatively young. It began to develop in the 70s of the XX century. During this time, work was carried out to restore certain territories disturbed by mining, construction work or brownfields for the sediments of industrial and household waste [5]. However, there is still no overall comparative concept of assessment of the environmental effectiveness of work carried out in different combinations of lithological and bioclimatic soil conditions. Still there is no conceptual justification for this direction from the point of view of the formation of sustainable nature-like ecosystems after reclamation. Along with data on the effectiveness of reclamation measures, the results of studies of tendencies and results of restoration processes will contribute to the creation of a holistic concept of regeneration of systems after destructive technological impacts [6, 7]. Terrestrial ecosystems normally are in dynamic equilibrium with environmental conditions, the biological cycle of elements in them is almost closed. Open mining completely destroys the soil and plant cover and biogeochemical cycles [8]. Self-restoration of zonal ecosystems takes 100 years and more. Reclamation techniques can accelerate the onset of individual stages of soil restoration by 10-30 years [9, 10]. Reclamation techniques developed in Russia and foreign countries are aimed mainly at restoring the organic part of disturbed ecosystems and initial soils. Biological reclamation (seeding of legumes, cereal-grass combinations, forest plantations) contributes to a more even development of the vegetation cover, which, in turn, accelerates the rate of accumulation of organic matter and biogenic elements in the soil [5]. The continuous increase in land areas disturbed by mining and low rates of natural soil regeneration results in formation of a disbalance of nature consumption and nature restoration. The increase in land areas affected by indirect mining and the frequent degradation of reclaimed land due to their low resilience in developing post-man-made landscapes become more and more intensive [6]. During restoring natural ecosystems in the process of self-thickening of dumps, the mechanisms of regeneration processes developed by nature for thousands of years are carried out, based on the biogenic development of the substrate through the accumulation and transformation of organic matter [11, 12]. These processes were well studied recently in terms of carbon stocks formation, humification rate, microbial biomass and respiration rate [5].

The classification of reclamation resources is presented in fig. 1. According to this ideology, one can see that microbiomes of mining heaps could be considered as one of the most important resources of reclamation.

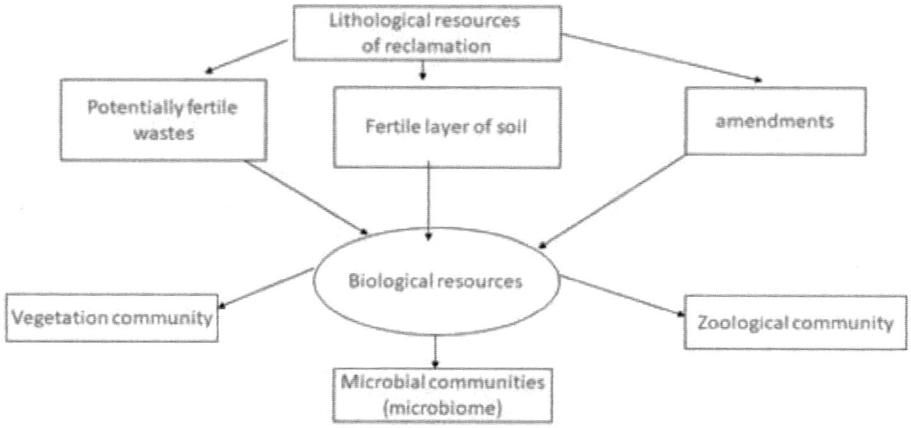

Figure 1. Classification of the reclamation resources

Nowadays, the development of microbial community on the stage of initial revegetation of mining wastes is underestimated in relevant studies. Thus, the aim of this study was to evaluate taxonomy composition of microbiome of initial soils and bio crusts of initially bio colonized mineral wastes of abandoned mines of Russian North-West.

2. Materials and methods

Various mining heaps materials has been sampled in field season of 2020 y. The samples were taken from various geoenvironments. Some of them were presented by quarternal mineral sediments, accumulated in mining heaps. Other ones were older in terms of ancient sedimentology and dealed chemical properties, but also exposed on modern surface for current biological influence. Field samples were transported in frozen and unfrozen from to laboratory. Air dried unfrozen samples were grounded and passed through 2 mm sieve. Routine soil analyses were performed according manual [13]. Photos of mining heap a given on the fig. 1.

Limestones crushing siftings

Granite crushing siftings

Carboniferous clays

Carboniferous clays

Quaternary sands

Toxic pyrite containing dumps

Figure 2. Examples of mining wastes studied

Frozen subsamples were used for metagenomic studies. DNA was isolated using the MN FastDNA Spin Kit (MN, Germany) using a Precellus 24 homogenizer (Bertin, USA). The quality control of the isolation was carried out by PCR and agarose gel electrophoresis. V4 sequencing of the variable region of the 16S rRNA gene was performed on an Illumina MiSEQ sequencer using primers F515 (GTGCCAGCMGCCGCGGTAA) and R806 (GGACTACVSGGGTATCTAAT) [14]. The general processing of sequences was carried out in the dada2 (v1.14.1) package [15], the main analysis of the results was carried out using the phyloseq (v1.30.0) package [16] in R (v3.6.3).

3. Results and discussion

The waste heaps studied were different in term of skeletal and fine earth fractions content which is normal for substrates of different origin [17, 18]. The fine earth of the mining wastes was quite diverse in terms of chemical indexes of fertility (table 1). The variations in TOC (total organic carbon), TON (total organic nitrogen), pH levels and granulometric (particle size) fractions groups indicated different origin and properties of the mineral wastes studied. In general, some of them suitable for intensive reclamation (quaternary sands) and some of them are not favourable due to chemical properties (toxic substrata).

Table 1. Texture indexes and chemical features of the mining wastes studied

Type of mineral waste	TOC, %	TON, %	pH	Skeletal fraction, %	Fine earth fraction, %
Limestones crushing siftings	1.34	0.15	7.8	73	27
Granite crushing siftings	0.53	0.04	6.7	67	23
Carbon clays	0.32	0.02	4.9	98	2
Quaternary sands	0.21	0.02	4.7	96	4
Toxic pyrite containing materials	0.05	0.01	2.8	76	24

The principal composition of samples microbiomes provided below. In most cases, top layer of self-growing sites are similar to their control soils in terms of taxonomical groups, but less diverse in alpha-diversity metrics. Key phyla are Proteobacteria, Actinobacteria, Acidobacteria, Bacteroidetes, Chloroflexia, Verrucomicrobia. This well corresponds for recently derived data on boreal and subboreal pedo environments [19-21]. In cambrian clay quarries there are numerous representatives of Verrucomicrobia, whereas for clay dumps there are range of autotrophic bacteria. Self-growing dumps of granite crushing siftings are characterised by essential presence of ammonium-oxidizing archaea (Nitrososphaeracea), probably, they can be caused by increased aeration. Refractory (carboniferous) clay quarries also provide toxing dumps communities are extremely low diverse and contain mainly Actinobacteria (up to 55%) and Chloroflexi (up to 15%). Revegetated heaps of carboniferous clays shows less Actinobacteria and more Cyanobacteria and Bacteroidetes, than in the control soils (Retisols or clayely textured podzolised soils of south taiga). Granite crushing siftings wastes shows less Acidobacteria and Actinobacteria. Normally, in non-toxic substrates with normal acidity main difference are in abundance on family, but, not phylum level, and these data should be compared with regional benchmark soils of the southern taiga ecoclimatic zone. The boreal coniferious forests formed on acid sands demonstrate roughly the same amount of Proteobacteria, but less Acidobacteria (decreases in older soils) and Actinobacteria (increases in older soils). Also there are more representatives of Bacteroidetes and less Verrucomicrobia. The forests soils communities are less variate in repeats, than disturbed mines, this well corresponds with some recent works [21, 22]. Data obtained should be incorporated into the strategy of wastes heaps reclamation in terms of quantification of possible microbial prerequisites of soil reclamation in diverse lithological and climatogenic conditions of soil restoration [23, 24].

Timing factor as well as lithological one plays an important role in taxonomical differentiation of microbiome. Thus, core composition is comparable on initial stages of biogenic-abiogenic interactions. Later, after 5-10 years of heap exposing on the surface of post mining site, differentiation of microbiome appears. This means that our current task is to collect data on soil microbiome quality and composition and organize these data to structured base. This base will allow to analyse combinations of lithological features of mining wastes and to select the decision for reclamation with use of key group of microorganisms.

4. Conclusions

The investigation conducted allows us to conclude that mining wastes are intensively colonized by various microorganisms even on every initial stages of their exposing on the surface of heaps. Thus, numerous representatives of bacteria's appears on their surface, which indicates the priming of biogenic-abiogenic interactions and forthcoming soil formation. Mining wastes are not lifeless sediments, accumulated on the surface of post technogenous landscapes, they accumulates various species of micriorganisms in superficial microlayers. Biolithogenic interactions on the surface of heaps could be considered as initial driver of self restoration and embryonic soil formation.

The first sketch of microbiological information, obtained in this study will be used in elaboration of further experimental designs of waste heaps reclamations in conditions of great variance of lithological conditions of fresh sediments exposures of Russian North-West. The priming of soil formation is evident, but its chemical, biological and biogeochemical mechanisms will be established in future researches, based on current data.

Acknowledgments
This work was supported Russian Science Foundation (project No 17-16-01030).

References

[1] Sokolov D, Morozov S, Abakumov E and Androkhanov V 2021 Polycyclic Aromatic Hydrocarbons in Soils of Anthracite Deposit Dumps in Siberia *Eurasian Soil Science* **54** 875–887
[2] Likhanova I and Archegova I 2009 Soil formation under progressive accelerated succession of forest ecosystems in the North of Russia *Contemporary Problems of Ecology* **2** 66–71
[3] Archegova I 2007 Thermal regime of tundra soils under reclamation and restoration of natural vegetation *Eurasian Soil Science* **40** 854–859
[4] Beyer L, Huyke W, Hüttmann S, Archegova I and Titarenko T V 2002 The Use of Microbial Activity Indicators for a Quality Assessment of Highly Crude Oil Contaminated Soils in the Russian Subpolar Tundra at the Arctic Circle *Polarforschung* **71** 33–39
[5] Lin Q, Dini-Andreote F, Li L, Umari R, Novotny V, Kukla J, Heděnec P and Frouz J 2021 Soil microbial interconnections along ecological restoration gradients of lowland forests after slash-and-burn agriculture *FEMS Microbiology Ecology* **97**
[6] Cortina J, García-Sánchez I, Grace M, Andrés P, Baker S, Bullock C, Decleer K, Dicks L, Fisher J, Frouz J, Klimkowska A, Kyriazopoulos A, Moreno Mateos D, Rodríguez-González P, Sarkki S and Ventocilla J 2021 Barriers to ecological restoration in Europe: expert perspectives *Restoration Ecology* **29**
[7] Mergner R, Janssen R, Rutz D, Malamatenios C, Veziryianni G, Knoche D, Rademacher A, Schlepphorst R, Davies S and Michie R 2020 Smart strategies for the transition in coal intensive regions *The 28th European Biomass Conference and Exhibition* 936–940
[8] Ivanova E A, Pershina E V, Shapkin V M, Kichko A A, Aksenova T S, Kimeklis A K, Gladkov G V, Zverev A O, Vasilyeva N A, Andronov E E and Abakumov E V 2020 Shifting prokaryotic communities along a soil formation chronosequence and across soil horizons in a South Taiga ecosystem *Pedobiologia* **81–82** 150650
[9] Hüblová L and Frouz J 2021 Contrasting effect of coniferous and broadleaf trees on soil carbon storage during reforestation of forest soils and afforestation of agricultural and post-mining soils *Journal of Environmental Management* **290** 112567
[10] Benetková P, Tichý L, Háněl L, Kukla J, Vicentini F and Frouz J 2020 The effect of soil and plant material transplants on vegetation and soil biota during forest restoration in a limestone quarry: A case study *Ecological Engineering* **158** 106039
[11] Moradi J, John K, Vicentini F, Veselá H, Vicena J, Ardestani M M and Frouz J 2020 Vertical distribution of soil fauna and microbial community under two contrasting post mining chronosequences: Sites reclaimed by alder plantation and unreclaimed regrowth *Global Ecology and Conservation* **23** e01165
[12] Kaneda S, Krištůfek V, Baldrian P, Malý S and Frouz J 2019 Changes in functional response of soil microbial community along chronosequence of spontaneous succession on post mining forest sites evaluated by Biolog and SIR methods *Forests* **10** 1005
[13] Bottomley P J, Angle J S and Weaver R 2020 *Methods of soil analysis.* Part 2: Microbiological and biochemical properties, Volume 12 (John Wiley & Sons)

[14] Caporaso J G, Lauber C L, Walters W A, Berg-Lyons D, Lozupone C A, Turnbaugh P J, Fierer N and Knight R 2011 Global patterns of 16S rRNA diversity at a depth of millions of sequences per sample *Proceedings of the National Academy of Sciences* **108** 4516

[15] Callahan B J, McMurdie P J, Rosen M J, Han A W, Johnson A J A and Holmes S P 2016 DADA2: High-resolution sample inference from Illumina amplicon data *Nature Methods* **13** 581–583

[16] McMurdie P J and Holmes S 2013 phyloseq: An R Package for Reproducible Interactive Analysis and Graphics of Microbiome Census Data *PLOS ONE* **8** e61217

[17] Ivanova E, Pershina E, Karpova D, Rogova O, Abakumov E and Andronov E 2019 Soil microbiome in chronosequence of spoil heaps of Kursk Magnetic Anomaly *Biological Communications* **64** 219–225

[18] Abakumov E, Zverev A, Kichko A, Kimeklis A and Andronov E 2021 Soil microbiome of different-aged stages of self-restoration of ecosystems on the mining heaps of limestone quarry (Elizavetino, Leningrad region) *Open Agriculture* **6** 57–66

[19] Abakumov E, Loyko S, Istigechev G, Kulemzina A, Lashchinskiy N, Andronov E and Lapidus A 2020 Soils of Chernevaya taiga of Western Siberia – Morphology, Agrochemical features, microbiome *Sel'skokhozyaistvennaya Biologiya [Agricultural Biology]* **55** 1018–1039

[20] Pershina E V, Ivanova E A, Korvigo I O, Chirak E L, Sergaliev N H, Abakumov E V, Provorov N A and Andronov E E 2018 Investigation of the core microbiome in main soil types from the East European plain *Science of The Total Environment* **631–632** 1421–1430

[21] Zverev A, Petrov A, Kimeklis A, Kichko A, Andronov E and Abakumov E 2020 Microbiomes of the initial soils of mining areas of Yakutsk City (Eastern Siberia, Russia) *Czech Polar Reports* **10** 69–82

[22] Kimeklis A, Dmitrakova Y A, Pershina E, Ivanova E, Zverev A, Gladkov G, Kichko A, Andronov E and Abakumov E 2020 Analysis of microbiome of recultivated soils of the Kingisepp area of phosphorite mining *Sel'skokhozyaistvennaya Biologiya* **55** 137–152

[23] Vodoleev A, Androkhanov V and Berdova O 2017 Yumashe va NA, Cherdantseva ES Environmentally safe storage of wastes from ironore enrichment *Izvestiya Vysshikh Uchebnykh Zavedenij. Chernaya metallurgiya* **60** 792–797

[24] Artamonova V, Androkhanov V, Sokolov D, Lyutykh I, Bulgakova V, Bortnikova S and Vodoleev A 2011 Ecological-physiological diversity of microbial communities in the anthropogenically impacted landscapes of Kuzbass *Contemporary problems of ecology* **4** 540–549

Waste from processing of technical grape varieties in poultry nutrition

Raisat Akhmedkhanova[1], Haji Shabanov[1], Samira Aliyeva[1], Albina Alakayeva[1], Irina Musayeva[1], and Patimat Hiramagomedova[1]

[1]Dagestan State Agrarian University named after M. M. Dzhambulatov RD, Russia, Makhachkala

E-mail: raisatragimovna@mail.ru

Abstract. The feed additives for balancing the diet for biologically active substances still utilizing up to the present time. Their main components are synthetic substances that have not only low digestibility, but also negatively affect human health. In this regard, we are developing technologies for the utilization of local plant feed additives and waste processing in the feeding of farm animals and poultry. This is the waste of processing technical grape varieties – grape pomace. These wastes can be considered as natural feed additives because after technological processing they can be used to produce flour with a high content of macro and microelements, vitamins, as well as amino acids.

1. Introduction

Particular attention has been focused recently on improving the quality of eggs and poultry meat by using various non-traditional natural feed additives. These feed additives based on natural plant raw materials are rich sources of biologically active substances of various actions. In addition, they are much cheaper due to the simplicity of harvesting and are environmentally friendly, so it is economically advantageous to introduce them into the diet for animals and poultry.

This is proved by many studies on the enrichment of the diet with non-traditional feed additives of plant origin, which help to raise both the taste qualities and the ecological purity of the produced products.

According to the data of Alieva S. M. et al. (2018), the replacing of herbal flour with nettle and seaweed flour in the feed of broiler chickens significantly affected the accumulation of iodine in the meat and liver of broiler chickens. [1,2]

Studies conducted by Akhmedkhanova R. R. et al. (2020) suggest the opportunity of obtaining iodized dairy products by enriching the diet of lactating cows with microalgae. [3,4]

Ponomarenko Yu. A. in his works notes the positive effect on the accumulation of iodine and selenium in the meat and liver of broiler chickens, depending on the amount of suspension and dry chlorella enriched with iodine. [5]

According to S. Leeson (2007), the production of poultry products enriched with vitamins leading the necessity for increased provision of poultry with vitamins through the feed. [6]

Khan R.U. (2012) suggests using ginger as feed additives in the diets of chickens to increase their potential productivity [7].

In their work, T. E. Abbas and M. E. Ahmed (2010) present a detailed review of the potential benefits of utilizing black cumin in the diets for broilers and laying hens. [8].

Valpotić H. in his study, notes the role of zeolites in balancing diets with minerals their positive effect on the animal body [9].

According to Ignatovich L. (2013), the use of cheap local natural feed additives from natural plant raw materials in poultry farming contributes to the production of environmentally friendly products [10].

Mayurnikova L. et al. (2020) suggests a technology for enriching plant raw materials with selenium and using it in the production of preventive food products and in animal feeding [11].

KiczorowskaB. (2016) notes that probiotics, prebiotics, and phytobiotics could be a possible solution as immunostimulants in monogastric animal nutrition [12].

Mutimura M. (2018) suggests that making use of legumes helpsto improve the quality of nutrition for lactating dairy cows[13].

Many researchers in their work noted that in the process of metabolism, and the biosynthesis of many nutrients and bioactive substances of the feed moving into the egg and muscle tissue, and in the human body [14-17].

Therefore, it can be noted that the use of natural feed additives helps to increase food production, improve their quality, biological value and sustainability. [18-21].

Previously, we studied and tested the waste of processing black technical grape varieties on broiler chickens and as a result, reliable indicators were obtained for improving the live weight and quality of broiler meat [22].

The purpose of the work is to study the waste of processing of white technical grape varieties (Rkatseteli) and their impact on egg production, and morphological indicators of the egg quality of laying hens the cross "Rhodonit 2" with the establishment optimal levels of their introduction into the diet.

2. Methods

The object of research was laying hens of the cross "Rhodonit 2" of the farm "Gadzhimurzayeva M. G.". For research, 3 groups of laying hens of 30 heads each at the age of 24 weeks were formed by the method of analog groups. They were fed with a full-fledged balanced compound feed (CF) in accordance with the norms of inter VNITIP [9]. The first control group received compound feed (CF) without the inclusion of flour from pomace, and 2 experimental CF with the inclusion of -3% flour from grape pomace and 3 experimental CF with the inclusion of 4%.

During the study period (20 weeks), egg production (gross egg collection for 150 days) was taken into account by daily counting the number of laid eggs with the further determination of the intensity of egg-laying (%) and the mass of egg products (kg) for the initial laying hen, as well as morphological and biochemical parameters of eggs, the content of carotenes, vitamins A, B2, vitamin C in the organs, heavy metals in the yolk and enzyme activity.

3. Results

As the research result, the following composition of flour from the pomace of white technical grape varieties was established (Table 1).

Table 1. Composition of flour from the pomace of white technical grape varieties (Rkatseteli).

№	Indicator	unit		№	Indicator	unit	
	Crude protein	%	13.79		Copper	mg/kg	6.7
	Crude fat	%	0.78		Manganes	mg/kg	8.9
	Crude fiber	%	23.4		Cobalt	mg/kg	0.14
	Calcium	g/kg	16		Zink	mg/kg	21
	Phosphorus	g/kg	2.61		Iron	mg/kg	174.5
	Carotene	mg/kg	0.46		Iodine (flour from pomace of white varieties (rkatseteli) grapes)	mg/kg	10.25
			Amino acids,%				
1	Lysine		0.40	11	Cystine		0.20
2	Histidine		0.28	12	Threonine		0.39
3	Serine		0.39	13	Glycine		0.71
4	Alanine		0.51	14	Isoleucine		0.46
5	Leucine		0.70	15	Phenylalanine		0.41
6	Methionine		0.28	16	Glutamic acid		1.34
7	Arginine		0.49	17	Aspartic acid		0.67
8	Proline		0.50		***The sum of amino acids of them:***		***8.43***
9	Valine		0.42		***irreplaceable***		***3.06***
10	Tyrosine		0.28		***replaceable***		***5.37***

When we compare the pomace from dark grape varieties the composition of flour from white varieties slightly prevails in terms of protein content, amino acids and trace elements.

As a result was established that the grape pomace flour affected positively the productivity and quality of laying hens' eggs.

Table 2. Productivity and quality of eggs

№	Indicator	Group		
		1 control	2 experimental	3experimental
	Age 44 weeks			
1	Gross collection of eggs (for 20 weeks)	2846	3038	2927
2	Egg production for the initial laying hen, pcs	18.97	20.25	19.51
3	Egg production rate, %	62.0	66.18	63.76
	Morphological parameters of eggs			
4	Egg mass, g	61.21 ± 0.80	**62.27**± 0.82	**62.11**± 0.40
5	Egg weight, kg (153 days)	174.20	189,18	181,79
6	Egg mass yield, kg / bird	5.80	6.31	6.06
7	Average feed consumption per 1 head per day, g	120	120	120
8	Feed consumption per 1 kg of egg mass, kg	3.09	2.85	2.97
9	Shell mass, g	7.18 ± 0.10	7.30 ± 0.12	7.25 ± 0.13
10	Yolk weight, g	18.30	18.52	18.41
11	Egg white, g	35.73	36.45	36.45
12	Ratio of egg white to yolk	1.98: 1	1.97:1	1.98:1
13	Shell thickness, mm	0.346 ± 0.04	0.351 ± 0.007	0.350 ± 0.003
14	Egg density, g / cm3	1.024 ± 0.02	1.042 ± 0.01	1.040 ± 0.02
15	Egg shape index,%	74.95 ± 0.80	75.**03** ±0.73	75.**06** ±0.59
	The biological value of eggs			
16	Carotenoids in the yolk, μg / g	19.0 ±0.49	21.**38**±0.53	**21.40**±0.71
17	Vitamin A in the yolk, μg / g	8.6 2.8	9.**35**± 3.0	9.**4** ±3.2
18	Vitamin B2 in the yolk, μg / g	5.21	**5.87**	5.**84**
	in the egg white	2.43	2.**70**	2.**84**
19	Iodine, mcg/100gr	170.09	**205.93**	**252.18**
20	Acid number of the yolk, mg	4.8	**5.0**	**5.0**
21	Content in the yolk, μg / 100g			
	Lead	1.95	0.**80**	**not found**
22	Cadmium	0.28	**0.12**	**trace**
23	Arsenic	**trace**	**not found**	**not found**

According to the number of laid eggs within 153 days of research, the second experimental group of chickens exceeded the control group by 6.75% and the third experimental group by 2.24%. Accordingly, the intensity of egg production is higher in the third experimental group of chickens. One of the most important indicators of egg productivity is the average mass of eggs. During the research period, an increase in the average egg mass in chickens of the experimental groups was noted by 1.06 g in the second and 0.9 g in the third experimental groups. Along with this, the feed costs per kg of egg mass in the experimental groups were reduced by 0.24kg and 0.12 kg or by 7.77% and 3.88%. In addition, the iodine in egg yolk content of the experimental groups was significantly higher compared to the control group: by 21% and 48.3% in the 2nd and the 3rd groups respectively.

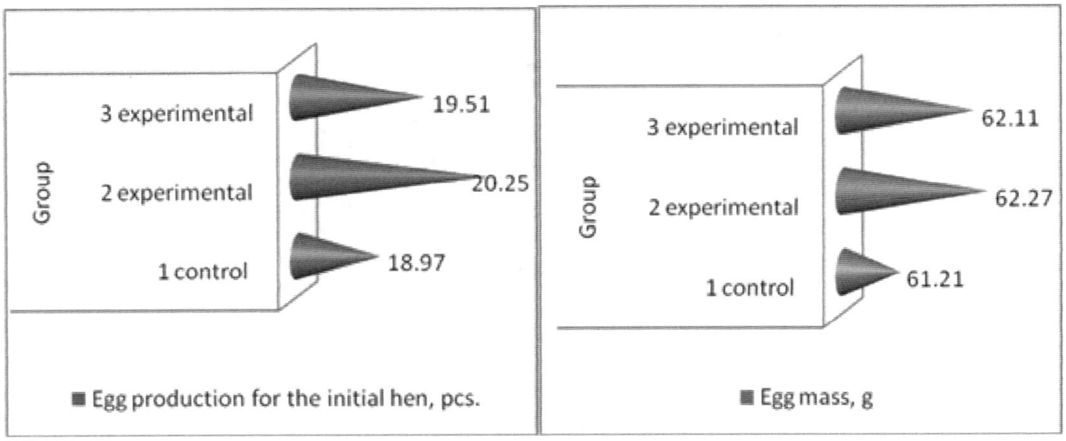

Figure 1. Egg production and weight of eggs

The thickness of the eggshell in the group of chickens that received grape processing waste was higher relatively to the control group by 1.45% and 1.16%. A similar pattern was observed for egg density.

Table 3. Vitamin C in organs and meat of laying hens receiving flour from grape pomace.

Indicators	Group		
	1 control	2 experimental	3 experimental
Liver	32.98 ± 2.58	37.22 ± 2.09[a]	40.09 ± 2.20[b]
Kidney	43.08 ± 1,92	46.28 ± 2.84	48,89 ± 1.89[a]
Heart	25.82 ± 1.82	28.88 ± 2.10	29,31 ± 2.91
Meat (white)	21.11 ± 3.12	29.06 ± 2.97[a]	30,15 ± 3.21[b]

[a]P<0,05.
[b]P<0,01.

The inclusion of grape processing waste in the diet of laying hens had a positive effect on the accumulation of vitamin C in some organs and in white (breast) meat (Table 3). In relation to the control group in all organs and meat, there is a tendency of increasing the accumulation of vitamins depending on the number of feed additives included in the diet from grape processing waste. The highest reliable (P<0.01) indicators for the accumulation of vitamin C were noted in the liver and in the white meat of laying hens that received 4% of grape pomace.

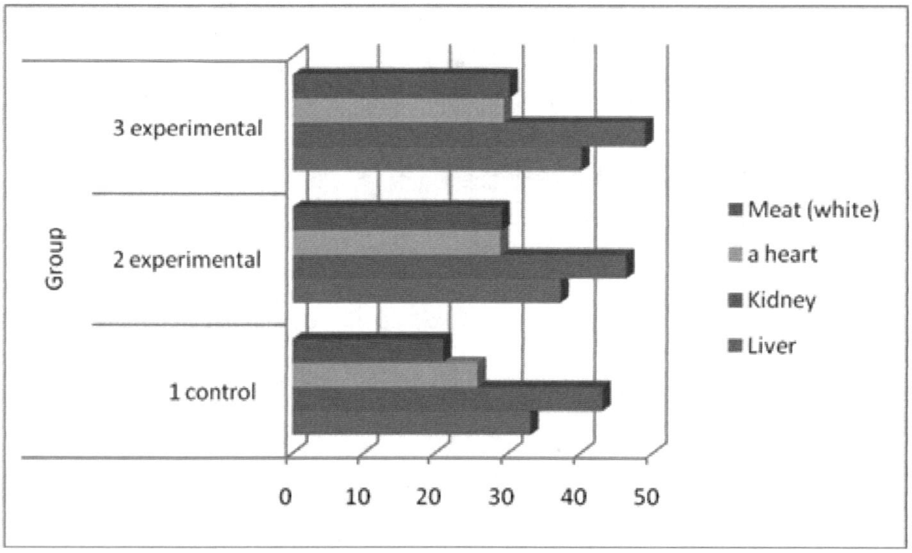

Figure 2. Vitamin C content

4. Discussion

In conclusion, the expediency of using waste from the processing of technical grape varieties as a feed additive for laying hens in the amount of 3%, since this improves egg production by 6.75%, reduces feed costs per 1 kg of egg mass by 7.76%, increases the thickness of the shell and the density of eggs, and also improves the biological value of eggs.

References

[1] Alieva S, Akhmedkhanova R and Astarkhanova T 2016 Local plant resources of the RD in the diet of broiler chickens *Scientific journal of KubSAU* **117(3)** 1–12

[2] Alieva S, Akhmedkhanova R, Gadzhaeva Z and Kozyrev S 2018 The effect of flour from stinging nettle and seaweed on the increase in the biological potential of the productivity of hens of the parent flock *Scientific and practical journal Izvestia of the Gorsk State Agrarian University* **55(2)** 64–67 (Vladikavkaz)

[3] Akhmedkhanova R, Alieva S, Gadzhaeva Z and Musakaeva S 2020 Iodization of livestock products using various feed additives of natural origin *All-Russian scientific-practical conference with international participation "Problems and prospects of organic agriculture development"* 208–213 (Makhachkala)

[4] Akhmedkhanova R, Dzhambulatov Z, Gadzhaeva Z, Shabanov G and Alieva S 2020 The influence of chlorella suspension on the quality of milk and its processing products *International Scientific and Practical Conference "Development of the Agro-Industrial Complex in the Context of Robotization and Digitalization of Production in Russia and Abroad" (DAIC 2020)* DOI: 10.1051/e3sconf/202022202021

[5] Ponomarenko Yu 2015 Chlorella enriched in iodine and selenium in the diets of laying hens Global Science and Innovation: materials of the *IV International Scientific Conference* **41** 15–20 (Chicago, 12–13 March 2015)

[6] Leeson S June 2007 Vitamin requirements: is there basis for re-evaluating dietary specifications *World's Poultry Science Journal* **63(2)** 255

[7] Khan R, Naz S and Nikousefat Z June 2012 Potential applications of ginger (Zingiberofficinale) in poultry diets *World's Poultry Science Journal* **68(2)** 245

[8] Abbas T and Ahmed M September 2010 The use of black cumin in poultry diets *World's Poultry Science Journal* **3** 519

[9] Valpotić H 2017 Zeolite clinoptilolitenanoporous feed additive for animals of veterinary importance: potentials and limitations *Periodicumbiologorum* **119(3)** 159–172

[10] Ignatovich L S 2014 Local plant resources in the diets of laying hens *Poultry and poultry products* **3** 32

[11] Mayurnikova L, Krapiva T, Kurakin M, Koksharov A and Shalimova D 2020 Technology of production of plant materials with high selenium content for food production *International Conference on Efficient Production and Processing (ICEPP-2020) EDP Sciences*, Internet Conferencing **161** 01032 5

[12] Kiczorowska B, SamolińskaW, Al-Yasiry A R M, Kiczorowski P and Winiarska-Mieczan A 2016 The natural feed additives as immunostimulants in monogastric animal nutrition – a review *Annals of Animal Science* DOI: 10.1515/aoas-2016-0076

[13] Mutimura M, Ebong C, Rao I M and Nsahlai I V 2018 Effects of supplementation of Brachiaria brizantha cv. Piatá and Napier grass with Desmodium distortum on feed intake, digesta kinetics and milk production in crossbred dairy cows *Animal Nutrition* **4(2)** 222–227

[14] Dyadichkin L, Kosintsev Y, Timofeeva E, Volchkov V, Yuchkina N and Padyukova N 2007 Comparative characteristics of eggs of chickens of egg crosses *Poultry and poultry products* **5** 41–43

[15] Kuznetsova T S 2007 Influence of biologically active additives on the quality of eggs *Poultry and poultry products* **1** 42–43

[16] Fisinin V, Shtele A and Erastov G 2008 The quality of food eggs and nutritional health *Poultry* **2** 2–6

[17] Krasnova O, Hardina E, Hramov S, Kudrin M and Konik N 2020 Ethological and biological features of the organism of the black-and-white bulls when using natural feed additives in the diets *BIO Web Conf.* **17** 00237 DOI: 10.1051/bioconf/20201700237

[18] Rusdy M 2018 A review of the effects of quality forage supplementation on growth performance of cattle and goats fed elephant grass basal diet *Livestock Research for Rural Development* **30(12)** 203

[19] Enkai Li, Jinbiao Zhao, Ling Liu and Shuai Zhang 2018 Digestible energy and metabolizable energy contents of konjac flour residues and ramie in growing pigs *Animal Nutrition* **4(2)** 228–233 DOI: 10.1016/j.aninu.2018.01.001 ISSN: 2405-6545

[20] Righi F, Pitino R, Manuelian C L, Simoni M, Quarantelli A, De Marchi M and Tsiplakou E 2021 Plant Feed Additives as Natural Alternatives to the Use of Synthetic Antioxidant Vitamins on Poultry Performances, Health, and Oxidative Status: A Review of the Literature in the Last 20 Years *Antioxidants* **10** 659 DOI: 10.3390/antiox10050659

[21] Yurina N, Khorin1 B, Yurin D, Semenenko M and Kuzminova E 2020 The effect of feeding a natural feed additive on the performance of broiler chickens XIII International Scientific and Practical Conference "State and Prospects for the Development of Agribusiness – INTERAGROMASH 2020" Edited by Rudoy D and Ignateva S. *E3S Web of Conferences* **175** id.04001 (Rostovon-Don, Russia) DOI: 10.1051/e3sconf/202017504001

[22] Akhmedkhanova R and Abdullabekov P 2012 *Waste of viticulture in feeding broiler chickens* State and prospects of innovative development of the agro-industrial complex: Sat. articles based on the materials of Mezhdunar. scientific-practical Conf., dedicated to the 5th anniversary of the Institute of Continuous Professional Education of Agroindustrial Complex Personnel of the Saratov State Agrarian University (N.I. Vavilova – Saratov) p 28–30

Biohydrogen production from food processing wastewater by a newly isolated thermophilic bacterium

Yuriy V Litti[1], Elena A Zhuravleva[1], Andrey A Kovalev[2], Dmitriy A Kovalev[2], Inna V Katraeva[3], and Sofiya N Parshina[1]

[1]Federal Research Centre "Fundamentals of Biotechnology" of the Russian Academy of Sciences
[2]Federal Scientific Agroengineering Center VIM
[3]Nizhny Novgorod State University of Architecture and Civil Engineering

E-mail: litty-yuriy@mail.ru

Abstract. The aim of this work was a comparative study of biohydrogen production from cheese whey and confectionary wastewater by a newly isolated thermophilic microbial strain *Thermoanaerobacterium thermosaccharolyticum* SP-H2. Experimental results showed that the fermentative hydrogen was successfully produced with the highest hydrogen yield of 3.9 mL H_2/mL cheese whey or 80 mL H_2/g chemical oxygen demand. The profile of soluble metabolite products showed that hydrogen generation by a new isolate was mainly acetate-type fermentation in the case of confectionary wastewater and mixed ethanol-acetate-lactate type fermentation in the case of cheese whey. The more optimal metabolic pathway of confectionary wastewater fermentation was confirmed by the better kinetic characteristics according to the Gompertz model.

1. Introduction

The rapid depletion of fossil fuels makes it important to find alternative energy sources that are environmentally friendly, cost-effective and sustainable. Currently, hydrogen is considered as an alternative environmentally friendly gas fuel of the future, to reduce dependence on fossil fuels and effectively initiating the energy transition [1]. Among biological methods of hydrogen production, dark fermentation has several advantages, such as higher hydrogen productivity, versatility, and high speed of the process. Many anaerobic organisms can produce hydrogen from waste. Of all hydrogen-producing microbes, mesophilic bacteria are the most widely studied, but thermophilic bacteria may be more promising due to certain operational advantages [2]. These advantages include: (1) higher rate and specific yield of hydrogen since high temperature accelerate biochemical reactions; (2) the possibility of direct use of high-temperature wastewater (WW) without pre-cooling [3]. Various mesophilic and thermophilic strains producing H_2 have been identified and extensively studied, including the *Clostridium*, *Enterobacter*, *Thermoanaerobacterium*, *Thermotoga* species, etc. These strains can differ significantly in the efficiency of the transformation of organic matter into biohydrogen. Isolation of new strains that effectively produce hydrogen is very important for the development of environmentally friendly energy [3].

The present study deals with new hydrogen-producing thermophilic microbial strain *Thermoanaerobacterium thermosaccharolyticum* SP-H2. Single strain cultivation was carried out to investigate the efficiency of hydrogen gas production from several food processing wastewaters, namely cheese whey and confectionery wastewater.

2. Materials and methods

2.1. Inoculum, substrates, mineral medium

The effluent from the thermophilic acidogenic reactor, inoculated with municipal sewage sludge and treating high-strength simulated wastewater [4], was used to isolate new bacterial strain.

Pfennig`s medium [5] containing the following (mg/l): NH_4Cl, 330; $MgCl_2 \cdot 6H_2O$, 500; $CaCl_2$, 168; KCl, 330; and KH_2PO_4, 330, was used to isolate a new bacterial strain. The medium was supplemented with yeast extract, 500 mg/L, and $NaHCO_3$, 2500 mg/L, as well as with a trace element solution (according to Lippert [6]) and a vitamin solution (according to Wolin [7]). Sodium sulfide (0.5 g/L) and cysteine (0.5 g/L) were used as reducing agents. The final pH was 6.8–7.0.

2 mL of effluent was inoculated into an 18 mL medium containing 5 g/L potato starch as the carbon source in a 60 mL serum bottle and cultured in anaerobic condition at 55°C. The enrichment culture was sequentially subcultured into a series of 5–10 serum bottles (dilution 10^{-6} - 10^{-11}), introducing 10% inoculum into each of them. From the last serum bottle, in which the growth and formation of hydrogen were recorded, reinoculation was repeated on a series of serum bottles. The genomic sequence of the isolate was deposited in the DDBJ/ENA/GenBank under the accession number NZ_VWOK00000000.1.

Several food processing wastewaters were used for conducting batch tests of isolated strain:

1) wastewater from closed joint stock company "Production Association "Gummi" (Lyskovo, Nizhny Novgorod, Russia), producing ingredients for the confectionery, dairy and bakery industries (hereafter, Gummi WW)

2) cheese whey from Pavlovsky Dairy Plant limited liability company (Pavlovo, Nizhny Novgorod, Russia) (hereafter, cheese whey).

Some parameters of Gummi WW and cheese whey are given in Table 1.

Table 1. Parameters of wastewater

Parameter	Units	Gummi WW	Cheese whey
pH	-	5.21	3.87
COD	mg O_2/L	19520	48984
TS	g/L	1.11	3.77
C/N	-	109.9/1	31.8/1
Fats	mg/L	14.2	102.8
Reducing sugar	%weight	0.613	n/a*
Non reducing sugar	%weight	0.127	n/a*
Total sugar	%weight	0.740	1.136

*n/a – not available

2.2. Conducting batch test

Batch experiments were carried out in 60 mL serum bottles with a working volume of ~20 mL containing 2 mL (10% v/v) of the strain SP-H2 suspension at exponential phase (OD600 = 0.8-1.0), 18 mL of Pfennig`s medium [5]. Gummi WW or Cheese whey was added as the carbon source. The volume of Gummi WW and Cheese whey added was 3.9 and 1.6 ml, respectively. The chemical oxygen demand (COD) of the added substrate in all bottles was 3.84 g O_2/L. The initial pH of the medium was adjusted to 7.0 with a 10% solution of HCl or NaOH. The serum bottles were closed with rubber stoppers and aluminium caps and purged with nitrogen gas to create anaerobic conditions. During the dark fermentation, biogas components and soluble metabolite products (SMPs) were monitored. Control serum bottles did not contain any carbon source, except endogenous carbon from cell suspension and some components (yeast extract, cysteine) of the medium. The hydrogen production in the control bottles was subtracted from the hydrogen production in treatment bottles for obtaining the actual hydrogen production from wastewater by SP-H2. Incubation was carried out at 55°C in the dark on a thermostated shaker with 100 rpm. All treatments were conducted in triplicates.

2.3. Analytical methods
The TS (Total Solids) weight was determined after drying the sample to constant weight at 105 °C. pH was measured with a pH meter FE20 (Mettler Toledo, Switzerland) equipped with InLab® Microelectrode (Mettler Toledo, Switzerland). The Carbon to Nitrogen (C/N) ratio was measured on an element analyzer Vario EL cube (Abacus Analytical Systems GmbH, Germany). The concentration of hydrogen, carbon dioxide and methane in the headspace was analyzed by a gas chromatograph Kristall 5000.2 (Chromatec, Russia), as described earlier [8]. SMPs in the liquid phase were analyzed by a gas chromatograph Kristall 5000.2 (Chromatec, Russia) (ethanol, acetate, propionate, n- and iso-butyrate, n- and iso-valerate, caproate) [9] and Kapel-205 high-performance capillary electrophoresis system (Lumex, Russia) (formate, succinate, lactate) [10]. Chemical oxygen demand (COD), fats and sugar were determined as described earlier [10].

2.4. Hydrogen production calculation, kinetic model
The amount of biogas in the serum bottles was determined indirectly by measuring the cumulative pressure in the headspace using a manometer. The pressure was converted to a volume of biogas at standard temperature and pressure according to the Mendeleev-Clapeyron equation for ideal gases. Cumulative hydrogen production was calculated from the headspace measurement of the gas composition and the total volume of biogas produced at each time interval using the mass balance equation [11]. Kinetic parameters of hydrogen production were determined using the modified Gompertz equation, as described elsewhere [11].

3. Results and discussion

3.1. Characterization of wastewater
The composition of the substrate is extremely important for obtaining biohydrogen. Hydrogen yield is primarily affected by the content of carbohydrates [10]. Some parameters of the wastewater used as substrates are presented in Table 1. It can be noted that the wastewater had a rather low pH, high COD and sugar content, especially cheese whey, which is considered a promising substrate for biohydrogen production [12-13].

Concerning the C/N ratio, cheese whey showed a value of 31.8/1 and Gummi WW – 109.9/1. Studies by other authors have shown that the correct C/N ratio significantly affects the production of biohydrogen by altering the metabolic pathways during dark fermentation [14]. A C/N ratio of 200 yielded the highest hydrogen yield and specific hydrogen production rate in the dark fermentation of wheat powder solution upon the addition of nitrogen [15]. An optimal C/N ratio of 137 to produce 3.5 mole H_2/mole sucrose was estimated by a mathematical approximation [16]. Lin and Lei [14] achieved an increase in hydrogen yield of 500% and 80% and a hydrogen production rate, respectively, at a C/N ratio of 47. C/N ratio of 21 was optimal in terms of the specific biohydrogen production rate and yield during co-digestion of cheese whey and vegetable fruit waste [17]. Although the optimal C/N values are in a fairly large range, however, Gummi WW had a more optimal value.

3.2. Biohydrogen and SMPs production
Figure 1 shows data on the hydrogen content in biogas released during dark fermentation of wastewater. During all fermentation assays, no methane was detected. The maximum content in biogas was observed when using Gummi WW and amounted to 33.3% at 26.5 h of incubation, and when using cheese whey – 29.7% at 46.5 h of incubation. Interestingly, the bacterial species *Thermoanaerobacterium thermosaccharolyticum* has not previously been used to produce hydrogen from confectionary wastewater and cheese whey. For comparison, when using a mixed culture, the hydrogen content in the biogas ranged from 10 to 54% reaching the maximum value at hydraulic retention time (HRT) of 4.5

days in an anaerobic filter (AF) and anaerobic fluidized bed (AFB), treating cheese whey [12]. A wide range of hydrogen concentrations (10.8–52.4%) in biogas was also observed by Marques et al. [18] during DF of cheese whey wastewater in AFB at different HRTs. High hydrogen content in biogas (60%) was observed by van Ginkel et al. using confectionary wastewater as substrate and heat-treated soil as inoculum [19].

Hydrogen yield (on a COD basis) with Gummi WW and cheese whey as substrate were approximately the same at about 80 mL H_2/g COD (Figure 2A). According to other authors, we have obtained average values for the yield of hydrogen. Thus, the hydrogen yield from cheese whey was significantly lower (up to 25 mL H_2/g COD) in [20]. At the same time, the hydrogen yield when using cheese whey was a maximum 180 mL H_2/g COD in the work of Mikheeva et al. [12]. 20 to 170 mL H_2/g COD was obtained from dark fermentation of confectionary wastewater by van Ginkel et al. [19].

Hydrogen yields per ml of wastewater varied significantly and amounted to 3.9 and 1.5 mL H_2 per ml cheese whey and confectionary wastewater, respectively (Figure 2B). The values we obtained were comparable to 0.1–2 ml H_2 per ml of confectionery wastewater obtained by van Ginkel et al. [19]. Using *Lactobacillus acidophilus* ATCC 4356 as inoculum, 1.66 mL H_2 per mL cheese whey was obtained in [21], which was much lower than values obtained in our work.

The main SMPs in dark fermentation of cheese whey were ethanol (10.8 mM), acetate (10.4 mM), and lactate (12.8 mM) (Table 2). A significant proportion of ethanol and lactate among SMPs indicated that hydrogen was not produced through the most optimal metabolic pathway [10, 22]. In contrast to cheese whey, the main SMP during DF of confectionary wastewater was acetate (21.6 mM or 51% of SMPs on mole basis), which indicates a more efficient use of this substrate by *Thermoanaerobacterium thermosaccharolyticum* SP-H2. Most likely, this difference in the composition of SMPs is related to the carbohydrate composition of these wastewaters. Probably strain SP-H2 was better at converting sucrose, glucose and fructose in confectionary wastewater to hydrogen than lactose, which was the main sugar in cheese whey.

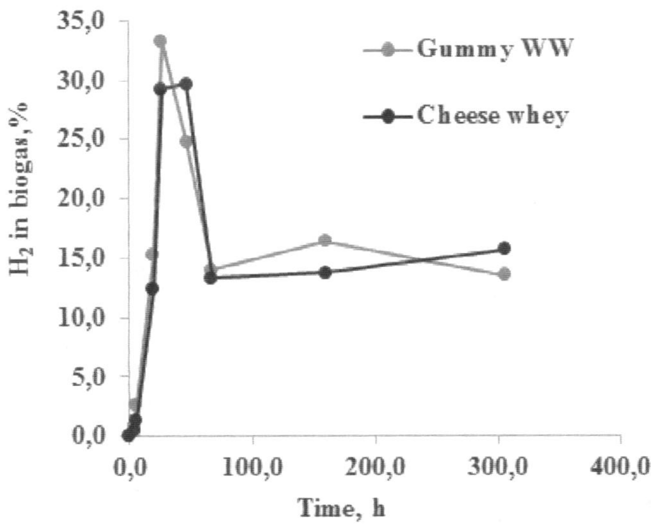

Figure 1. Hydrogen content in biogas.

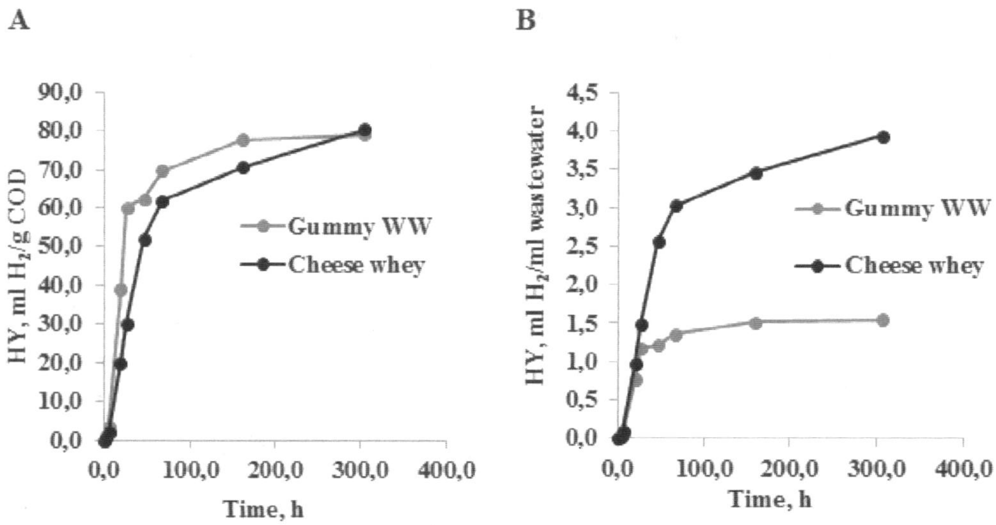

Figure 2. Hydrogen yield per g COD (A) and per ml (based on wastewater at its original strength (not diluted)) wastewater (B)

Table 2. pH and the main soluble metabolite products at the end of the batch test

Parameter	Units	Gummy WW	Cheese whey
pH	-	5.67	5.48
Ethanol	mM	10.1	10.8
Acetate	mM	21.6	10.4
Butyrate	mM	3.5	1.3
Formate	mM	1.6	0.0
Succinate	mM	0.1	0.1
Lactate	mM	5.1	12.8

3.3. Kinetic parameters

The kinetic parameters derived from the fitting of the experimental data with the Gompertz equation are given in Table 3. The Gompertz model fitted the experimental data well, with high coefficients of determination for all the assays (R^2 = 0.99). Potential hydrogen yield γ was slightly lower than experimentally obtained, while the γ value was slightly higher for cheese whey. The lag phase λ was practically the same for both wastewaters, while the maximum hydrogen production rate K was 2.4 times higher in the case of Gummy WW. The time needed to attain 90% of the potential hydrogen yield (t_{90}) and the effective time t_{eff} (calculated as the difference between t_{90} and lag phase λ) were 2.15–2.5 times lower when using Gummy WW as substrate.

Table 3. Kinetic parameters for hydrogen production from wastewater

Kinetic parameter	Units	Gummy WW	Cheese whey
γ	mL H_2/g COD	71.2	74.4
K	mL H_2/g COD/h	3.31	1.39
λ	h	6.97	6.36
t_{90}	h	32.6	70.4
t_{eff}	h	25.7	64.0
R^2	-	0.99	0.99

4. Conclusion

This study was a basic trial to evaluate the feasibility of producing hydrogen by the new thermophilic strain *Thermoanaerobacterium thermosaccharolyticum* SP-H2 in the dark fermentation process. Experimental studies revealed that this isolate successfully produced fermentative hydrogen from cheese whey and confectionery wastewater with the highest hydrogen yield of 80 mL H_2/g COD. DF of confectionery wastewater was distinguished by a higher maximum hydrogen production rate and less time needed to attain the potential hydrogen yield. The fast onset of hydrogen generation following medium inoculation makes *Thermoanaerobacterium thermosaccharolyticum* SP-H2 a promising strain for further study to enhance hydrogen yield.

Acknowledgments

This work was partially supported by the Federal State Budgetary Institution "Russian Foundation for Basic Research" as part of scientific project No. 18-29-25042.

References

[1] Wu X Y, Luo Y, Hess F and Lipinski W 2021 Sustainable Hydrogen for Energy, Fuel and Commodity Applications *Front. Energy Res.* **9** 231 DOI: 10.3389/fenrg.2021.698669

[2] Pawar S S and van Niel E W 2013 Thermophilic biohydrogen production: how far are we? *Appl. Microbiol. Biotechnol.* **97**(18) 7999–8009 DOI: 10.1007/s00253-013-5141-1

[3] Łukajtis R, Hołowacz I, Kucharska K, Glinka M, Rybarczyk P, Przyjazny A and Kamiński M 2018 Hydrogen production from biomass using dark fermentation *Renew. Sustain. Energy Rev.* **91** 665–694 DOI: 10.1016/j.rser.2018.04.043

[4] Kovalev A A, Kovalev D A, Litti Y V and Katraeva I V 2020 Biohydrogen production in the two-stage process of anaerobic bioconversion of organic matter of liquid organic waste with recirculation of digister effluent *Int. J. Hydrog. Energy* **45(51)** 26831–26839 DOI: 10.1016/j.ijhydene.2020.07.124

[5] Pfennig N 1965 Anreicherungskulturen fürote und grüne Schwefelbakterien *Zbl. Bakt. I Abt. Orig. Suppl.* **1** 179–189

[6] Pfenning N and Lippert K D 1966 Über das Vitamin B12 Bedürfnis phototropher Schwefelbakterien *Arch. Microbiol.* **55** 245–246 DOI: 10.1007/BF00410246

[7] Wolin E A , Wolin M J and Wolfe R S 1963 Formation of Methane by Bacterial Extracts *J. Biol. Chem.* **238** 2882–2886 DOI: 10.1016/S0021-9258(18)67912-8

[8] Lokshina L, Vavilin V, Litti Y, Glagolev M, Sabrekov A, Kotsyurbenko O and Kozlova M 2019 Methane Production in a West Siberian Eutrophic Fen Is Much Higher than Carbon Dioxide Production: Incubation of Peat Samples, Stoichiometry, Stable Isotope Dynamics, Modeling *Water Resour* **46(S1)** 110–115 DOI: 10.1134/S0097807819070133

[9] Litti Y, Nikitina A, Kovalev D, Ermoshin A, Mahajan R, Gunjan G and Nozhevnikova A 2017 Influence of cationic polyacrilamide flocculant onhigh-solids' anaerobic digestion of sewage sludge under thermophilic conditions *Environ Technol.* **14** 1–26 DOI: 10.1080/09593330.2017.1417492

[10] Litti Y V, Kovalev A A, Kovalev D A, Katraeva I V, Parshina S N, Zhuravleva E A and Botchkova E A 2021 Characteristics of the process of biohydrogen production from simple and complex substrates with different biopolymer composition *Int. J. Hydrog. Energy* DOI: 10.1016/j.ijhydene.2021.05.165

[11] Litti Yu V, Kovalev D A, Kovalev A A, Katraeva I V, Mikheeva E R and Nozhevnikova A N 2019 Use of a vortex layer apparatus for improving the efficiency of methane digestion of wastewater sludge *Water supply and sanitary technique* **11** 32–40 DOI: 10.35776/MNP.2019.11.05

[12] Mikheeva E R, Katraeva I V, Kovalev A A, Kovalev D A, Nozhevnikova A N, Panchenko V, Fiore U and Litti Y V 2021 The Start-Up of Continuous Biohydrogen Production from Cheese Whey: Comparison of Inoculum Pretreatment Methods and Reactors with Moving and Fixed

Polyurethane Carriers *Appl. Sci.* **11(2)** 510 DOI: 10.3390/app11020510

[13] Rao R and Basak N 2021 Fermentative molecular biohydrogen production from cheese whey: present prospects and future strategy *Appl. Biochem. Biotechnol.* 1–34 DOI: 10.1007/s12010-021-03528-6

[14] Lin C Y and Lay C H 2004 Carbon/nitrogen-ratio effect on fermentative hydrogen production by mixed microflora *Int. J. Hydrog. Energy* **29** 41–45 DOI: 10.1016/S0360-3199(03)00083-1

[15] Argun H, Kargi F, Kapdan I K and Oztekin R 2008 Biohydrogen production by dark fermentation of wheat powder solution: Effects of C/N and C/P ratio on hydrogen yield and formation rate *Int. J. Hydrog. Energy* **33** 1813–1819 DOI: 10.1016/j.ijhydene.2008.01.038

[16] del Pilar Anzola-Rojas M, da Fonseca S G, da Silva C C, de Oliveira V M and Zaiat M 2015 The use of the carbon/nitrogen ratio and specific organic loading rate as tools for improving biohydrogen production in fixed-bed reactors *Biotechnol. Rep.* **5** 46–54 DOI: 10.1016/j.btre.2014.10.010

[17] Gomez-Romero J, Gonzalez-Garcia A, Chairez I, Torres L and Garcia-Peña E I 2014 Selective adaptation of an anaerobic microbial community: Biohydrogen production by co-digestion of cheese whey and vegetables fruit waste *Int. J. Hydrog. Energy* **39(24)** 12541–12550 DOI: 10.1016/j.ijhydene.2014.06.050

[18] Marques T D, Macêdo W V, Peiter F S, Bonfim A A, Sakamoto I K, Caffaro Filho R A, Marcia H Z Damianovic, Varesche M B A, Salomon K R and de Amorim E L 2019 Influence of hydraulic retention time on hydrogen production by treating cheese whey wastewater in anaerobic fluidized bed bioreactor-an approach for developing countries *Braz. J. Chem. Eng.* **36(3)** 1109–1117 DOI: 10.1590/0104-6632.20190363s20190075

[19] van Ginkel S W, Oh S E and Logan B E 2005 Biohydrogen gas production from food processing and domestic wastewaters *Int. J. Hydrog. Energy* **30(15)** 1535–1542 DOI: 10.1016/j.ijhydene.2004.09.017

[20] Fernández C, Carracedo B, Martínez E J, Gómez X and Morán A 2013 Application of a packed bed reactor for the production of hydrogen from cheese whey permeate: Effect of organic loading rate *J. Environ. Sci. Health Part A* **49** 210–217 DOI: 10.1080/10934529.2013.838885

[21] Pandey A, Srivastava S, Rai P and Duke M 2019 Cheese whey to biohydrogen and useful organic acids: a non-pathogenic microbial treatment by L. acidophilus *Sci. Rep.* **9** 1–9 DOI: 10.1038/s41598-019-42752-3

[22] Ghimire A, Trably E, Frunzo L, Pirozzi F, Lens P N, Esposito G, Cazier E A and Escudié R 2018 Effect of total solids content on biohydrogen production and lactic acid accumulation during dark fermentation of organic waste biomass *Bioresour. Technol.* **248** 180–186 DOI: 10.1016/j.biortech.2017.07.062

IOP Conf. Series: Earth and Environmental Science **938** (2021) 012018 doi:10.1088/1755-1315/938/1/012018

Comparative evaluation of environmental efficiency of the municipal solid waste treatment scenarios for St. Petersburg region with the life cycle assessment method

N V Mikhailova[1] and A V Yasinskaya[1]

[1]REC "Mekhanobr-tekhnika", St. Petersburg, 22nd line, 3, Russia

E-mail: mikhailofff@gmail.com

Abstract. The research focuses on complex comparative evaluation of environmental efficiency of Municipal Solid Waste (MSW) treatment scenarios that can be implemented in St. Petersburg and Leningradskaya oblast'. Three scenarios were modelled that technically and economically meet the regional needs. The scenarios imply the following systems: 1) sorting, Mechanical-Biological Treatment (MBT), landfilling; 2) sorting, MBT, Waste-to-Energy (WTE), landfilling; 3) sorting, MBT, Refuse Derived Fuel (RDF) for cement industry, landfilling. In comparison, the "zero" scenario implying landfilling untreated waste was modelled. A Life Cycle Assessment based approach was proposed, which allows comparing scenarios through converting heterogeneous impact categories to a single monetary index. The highest damage reduction (by 60-70% in comparison with the "zero" scenario) was shown in the complex scenarios, including both MBT and energy recovery. 40, 42 and 47% of the total damage in scenarios 1, 2, 3 accordingly is associated with MBT product landfilling. The results show that the main benefit of complex treatment cannot be fully realized without separate collection of food waste. In this situation the most environmentally soundable scenario comprises recyclables separation and subsequent WTE treatment of all the rest with the exception of some non-combustibles.

1. Introduction

Choosing municipal solid waste (MSW) processing technologies is an urgent task in development of a waste management system. The task is to be addressed through analysis of a set of parameters including environmental safety. It is environmental safety that is the cornerstone of decision-making, which causes the strongest confrontation among the professionals and the public. Any party with an opposing interest easily steers the discussion towards emotional and logical speculations and blocks decision-making. This tends to happen already after immense sums have been spent on engineering and project design.

This issue is especially up to date for St. Petersburg region, where no new treatment facilities have been built since 1994, whereas the old ones have been shut down without the prospect of restarting. At the moment, the region of St. Petersburg – Leningradskaya oblast' has to build a new ecologically safe system for processing and disposal of 2,386,505 tons MSW annually [1, 2].

The research is dedicated to complex comparative evaluation of environmental efficiency of MSW treatment technological scenarios that can be implemented in the region of St. Petersburg – Leningradskaya oblast'. A Life Cycle Assessment based approach was proposed that allows to compare scenarios through converting heterogeneous impact categories to a single monetary index.

IOP Conf. Series: Earth and Environmental Science **938** (2021) 012018 doi:10.1088/1755-1315/938/1/012018

2. Materials and methods

Waste treatment scenarios were modelled that technically and economically meet the regional needs. The scenarios are based on technologies recommended for the Russian Federation [3, 4]. These scenarios also comply with the regional infrastructure and allow meeting the goals of the Federal Project "Developing the complex system for MSW treatment" (a part of the National Project "Ecology") and the Presidential Decree № 474 from 2020, July 7th. The scenarios description is presented in Table 1 and Figures 1-3.

In comparison, the "zero" scenario implying continued landfilling untreated waste was modelled. This scenario includes construction and operation of sanitary landfills only.

Table 1. Description of the modelled scenarios

Scenario	Technologies	Processing products handling
Scenario 0	– Landfilling non-stabilized wastes	Landfilling non-stabilized wastes
Scenario 1	– Sorting, – Mechanical Biological Treatment (MBT), – Landfilling stabilized wastes	Recyclables processing. Landfilling inerts and MBT product.
Scenario 2	– Sorting, – Mechanical Biological Treatment (MBT), – Landfilling stabilized wastes, – Energy recovery: Waste-to-Energy for combustibles (WtE), – Landfilling stabilized wastes.	Recyclables processing. Landfilling inerts and MBT product. Landfilling immobilized ash and slag. Electricity to the grid.
Scenario 3	– Sorting, – Mechanical Biological Treatment (MBT), – Landfilling stabilized wastes, – Energy recovery: Refuse Derived Fuel from sorting residues (RDF), – Landfilling stabilized wastes.	Recyclables processing. Landfilling inerts and MBT product. Refuse Derived Fuel for fossil fuel substitution in the cement plants of Leningradskaya oblast' and other regions

Modelling was fulfilled using the Life Cycle Assessment method (LCA). The model describes environmental impact during the full life cycle of MSW to the moment of final stabilization at a landfill site. Quantitative description of material and energy transformation by means of Material Flow Analysis is an important feature of the model.

For modelling, each technology was presented as a combination of elementary processes, connected with each other by elementary material and energy flows. The system boundaries involve sorting, residual and biodegradable waste processing with thermal and biological treatment, refuse derived fuel (RDF) use in cement industry, landfilling processed waste. Energy and resource effectiveness was considered indirectly by adding a carbon footprint of technologies into the model.

Since the research aim lies within the scope of comparative evaluation, there were chosen and analyzed only differing life cycle stages and impact aspects. In this research, the quantity of transport work was not included into modelling: all scenarios imply using the same facility sites, while the landfills are situated at the same sites with the treatment. However, the developed algorithm allows considering transportation as well.

The preparatory stage of modelling included:

- the characterization of the technologies,
- MFA description of quantitative and qualitative material and energy transformations in the processes under analysis,
- analysis of routes of substances emission and migration to the environment.

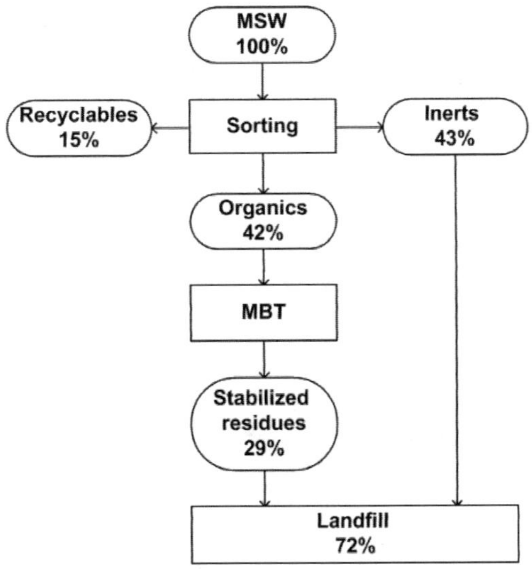

Figure 1. Material flow diagram of Scenario 1. Sorting, MBT, landfilling inert and stabilized waste

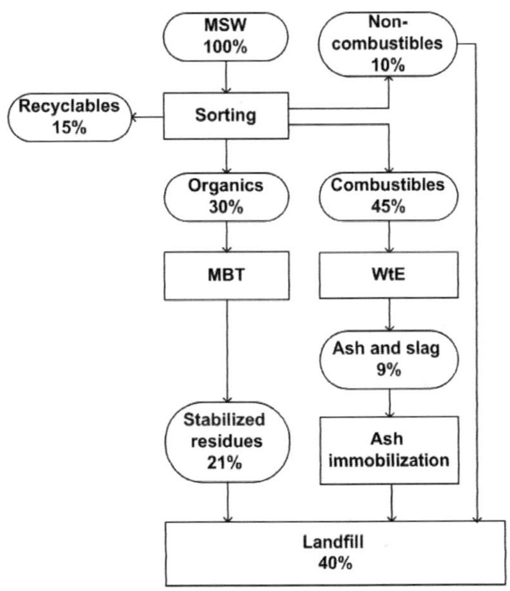

Figure 2. Material flow diagram of Scenario 2. Sorting, MBT, WTE, landfilling inert and stabilized waste

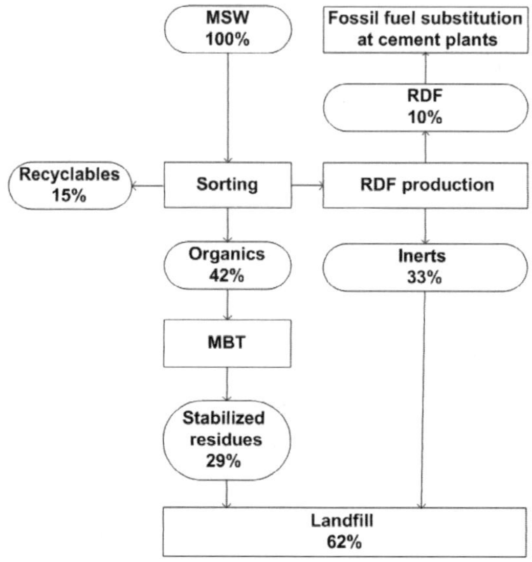

Figure 3. Material flow diagram of Scenario 3.

The research database relies on over 200 documents: statistics, open databases, books, standards, guides, industrial control and expertise data, reports, scientific papers. The dataset collected was systematized, all data were converted to required numerical figures that can be used in modelling.

The identified factors of impact were sorted out into environmental categories — landscape degradation and withdrawn soils from circulation, atmosphere pollution with toxicants, greenhouse gases emission, hydrosphere pollution, soil pollution, thermal pollution, noise and vibration, electromagnetic fields, ecosystems degradation, energy and material efficiency.

Within each category the analysis of impact mechanisms and engaged flows was made. At this and at the following stages there was conducted the screening of processes, flows and categories to be included into the model. The decision-making here was guided by a system of exclusion criteria: an impact comparison with the total impact in the system, environmental significance estimation, possibility of quantitative evaluation, a fact of identity for all scenarios, avoidance of double counting. The following impact categories were chosen for assessment: atmosphere pollution with toxicants, hydrosphere pollution, landscape degradation and withdrawn soils from circulation, soil pollution, energy and carbon efficiency (through carbon footprint).

In order to provide consistency in results, conditional industrial complexes with an identical capacity were considered as model objects. The model capacity is defined by technologies that should have a large scale for sufficient economic efficiency, namely landfills and Waste-to-Energy facilities (WTE). The recommended capacity confirmed by EU practice for such objects is about 500,000 ton/year [2]. Sorting, Mechanical-Biological Treatment (MBT), RDF production can work at the same or less capacity. Mass flow balances and important characteristics were assigned to the model facilities.

A model landfill complies with the environmental regulation of Russia. The basement layer comprises a membrane with permeability according to the SanPiN 2.1.3684-2021. Leachate is being collected and processed. The landfill body is covered with soil, landfill gas is being collected and flared. The difference between emissions from different treatment products was taken into account [5-10].

Next, characteristic models describing environmental impact were developed for each technology. Characteristic models describe functioning of technologies through corresponding elementary processes. The elementary processes are shown in Table 2.

Table 2. The elementary processes considered.

Technology		Elementary processes
Landfilling	–	Landfill gas flaring
	–	Emissions of uncollected landfill gas
	–	Transportation on site, loading and unloading, other support operations
	–	Leachate leakages through undamaged membrane lyners
	–	Leachate leakages through defects in membrane lyners
	–	Use of territory (soils withdrawal)
	–	Soil pollution because of waste deposition
	–	Soil pollution because of fall-out from the atmosphere.
Industrial processing	–	Sorting facility emissions
	–	MBT emissions (composting and maturation stages)
	–	Incineration emissions (WTE facility)
	–	Use of territory (soils withdrawal)
	–	Additional emissions due to RDF use at a cement plant
	–	Soil pollution around a WTE facility because of fall-out from the atmosphere.
	–	Soil pollution around a cement plant because of fall-out from the atmosphere
	–	Energy balance
	–	Landfilling processed residues (see above elementary processes for "Landfill")

The results of modelling have been received in the form of numerical values characterizing the impact factors (including positive effects). In order to provide comparability of results, the values of heterogeneous factors should be normalized, i.e. expressed in uniform units, for example in monetary units of environmental damage. Two methods are available for this purpose: a prevention based Eco-Costs method [11, 12] and a damage based "Temporary Method for estimation of prevented environmental damage" developed in Russia [13]. The calculations are relatively easy and transparent; the results are explained in terms of money and in measures to be taken.

The "Temporary Method..." has been chosen as the most appropriate for the research goals. It implies the above-mentioned impact categories and sets a cost for each factor by covering the whole complex of consequences for nature and humans. The method does not contain information about greenhouse emissions costs, so the carbon footprint was estimated according to the Eco-cost monetary values that were indexed using the average index for the Russian Federation from Eco-cost tables.

The model describes specific damage per ton of MSW entering the system. Estimation on a regional scope can be completed through multiplying the specific damage by the volume supposed to be treated.

3. Results and Discussion

The modelling results are summarized in Tables 3, 4.

Table 3. The results of modelling MSW treatment scenarios in terms of environmental damage.

Category of impact	Specific damage, $ per ton of MSW			
	Scenario 0 Landfilling	Scenario 1 Sorting, MBT, landfilling	Scenario 2 Sorting, MBT, WtE, landfilling	Scenario 3 Sorting, MBT, RDF, landfilling
1. Atmosphere pollution	45.87	0.04	0.07	0.03
2. Hydrosphere pollution	6.38	1.59	0.46	0.62
3. Soils withdrawn	73.66	55.58	30.92	40.72
4. Soil pollution	262.69	189.14	96.43	141.85
5. Carbon footprint	9.65	1.36	-6.43	-19.35
Total damage	**398.25**	**247.70**	**121.46**	**163.87**
Scenario cost (capex + opex)	**10.14**	**45.14**	**62.50**	**46.11**
Scenario cost + Damage	**408.39**	**297.35**	**190.21**	**214.59**
	Total annual damage for St.Petersburg and Leningradskaya oblast', $			
Total damage	**950 000 000**	**591 000 000**	**290 000 000**	**391 000 000**

Table 4. Prevented damage in comparison to the "zero" scenario

Category of impact	Damage	Damage prevented in comparison to "Zero" scenario		
	Scenario 0 Landfilling	Scenario 1 Sorting, MBT, landfilling	Scenario 2 Sorting, MBT, WtE, landfilling	Scenario 3 Sorting, MBT, RDF, landfilling
1. Atmosphere pollution, $/t	45.9	-45.8	-45.8	-45.8
2. Hydrosphere pollution, $/t	6.4	-4.8	-5.9	-5.8
3. Soils withdrawn, $/t	73.7	-18.1	-42.7	-32.9
4. Soil pollution, $/t	262.7	-73.6	-166.3	-120.8
5. Carbon footprint, $/t	9.7	-8.3	-16.1	-29.0
Total, $/t	**398.3**	**-150.6**	**-276.8**	**-234.4**
Total annual damage for the region, $	**950 000 000**	**-359 000 000**	**-661 000 000**	**-559 000 000**

It is important to consider the characteristics of the MSW collection system for correct assessment of the scenarios described. At the moment, the region has a commingled collection system. Separate collection of hazardous waste is organized unsatisfactorily.

In the future there is every possibility of the "Single Stream" collection system (all recyclables together), with elements of the multistream collection of recyclables in some collection points. There is no political willingness for organizing a separate collection of pure biodegradable organics suitable for producing marketable compost.

Accordingly, a scenario with composting for a marketable product cannot be implemented. The only possible direction for use of the "compost" from commingled waste is covering waste at landfill sites, which is equal to landfilling MBT product. This exact situation is described in scenarios 1-3.

"Zero" scenario: Sanitary landfill. The scenario demonstrates the highest impact on the environment in all categories considered. The most dramatic difference is the atmosphere pollution which is three orders of magnitude higher than in the other scenarios. This aspect of impact is connected with uncontrollable emissions that practically constitute about 70% of the total landfill gases generation [14-17]. Flaring collected landfill gas contributes much less into atmospheric emissions [18, 19].

The scenario demonstrates substantial differences in the hydrosphere impact. In comparison with the complex scenarios 2 and 3, the pollution in the "zero" scenario is one order of magnitude higher. The most important factor here are leakages through geomembrane defects which inevitably emerge during landfill construction and operation [20, 21].

The greatest contribution to the total environmental impact is from soil pollution affiliated both with direct pollution from waste deposition and fall-out from the atmosphere.

The scenarios designed to mitigate this impact are discussed further.

Scenario 1. Sorting, MBT, Landfilling. The scenario implying Mechanical Biological Treatment as the only waste stabilisation method has the least capacity to decrease the impact. The reason is a great amount of landfilled waste that, when stabilized, still is a source of emissions. The MBT process itself causes comparatively small impact (19% of the total impact in the system).

As a result, the MBT scenario allows reducing the damage by 40% in comparison with the "zero" scenario, that, certainly, justifies the cost incurred.

Scenario 2. Sorting, MBT, WTE, Landfilling. From all scenarios that were modelled, Scenario 2 demonstrates the highest environmental efficiency. This is facilitated by a significant decrease of mass in thermal treatment and by ash immobilization.

In spite of the fact that, from all treatment methods, WTE causes the highest atmosphere pollution (1.75 and 2.3 times more than scenarios 1 and 3), this damage is minor in comparison with sanitary landfill emissions to the atmosphere.

The shares of contributions to soil pollution are different in this scenario. The soil pollution around WTE facility is added, while the soil pollution associated with landfill is reduced. Overall, the soil pollution is decreased by comparison with the other scenarios.

As a result, the complex scenario including MBT and WTE allows to maximally reduce the environmental damage which is 70% lower than in the "zero" scenario.

Scenario 3. Sorting, MBT, RDF, Landfilling. The complex scenario including MBT, RDF production from sorting inerts, RDF use in cement kilns allows decreasing the damage by 60% in comparison with the "zero" scenario. In the circumstances described, Scenario 3 is slightly inferior to Scenario 2 in ecological terms, but shows an improvement compared to Scenario 1.

RDF use in cement kilns per se is the ecologically safest method for energy recuperation. Due to small recovery of RDF, a significant volume of inerts is being landfilled, which defines the main negative impact in Scenario 3. A quantity of recovered RDF technically can be increased to 18% in the scenario mass balance (due to inerts) or to 40% (due to inerts and dried organics) [22, 23, 24]. However, in this case the volume of marketable RDF will exceed the theoretical demand of cement plants in Leningradskaya oblast', and we have to consider more remote consumers that are potentially provided with RDF on the short arm (for example, from Moskovskaya and Kaluzhskaya oblasts' MSW processing facilities). In that scenario, the financial risks will increase while the ecological performance will improve.

All scenarios. Analyzing the scenarios modelled, we can conclude that industrial processing provides huge opportunities for decreasing atmosphere emission and achieving carbon neutrality. Atmosphere pollution reduction exceeded 99% for all scenarios in comparison with untreated waste landfilling.

Potential for decreasing the impact on the hydrosphere is limited by the possibility of decreasing landfilling volumes. The main source of hydrosphere pollution is deposit sites, and more specifically, leachate leakage through defects of lyners. The use of more expensive double composite lyners will reduce the impact a few times, but not eliminate it [25, 26].

The greatest contribution to the total environmental impact is from soil pollution, which is associated both with direct pollution from waste deposition and fall-out from the atmosphere. In scenarios implying energy recuperation (2 and 3) the damage to soils decreases 1.4 and 2.7 times, while the thermal processes itself cause only 26% and 20% of total soil pollution.

The lowest potential in impact decrease due to a system improvement lies in the withdrawn soils from circulation, which is reasonably connected with the size of facilities sites.

A substantial share of impact in scenarios 1-3 is associated with MBT product landfilling (40, 42 and 47% of the total damage in a scenario). This footprint is meaningful in every category, except the atmosphere pollution. The scenarios would have become more ecologically efficient if it was possible to add manufacturing marketable compost, as the main benefit of complex treatment is the opportunity to process each fraction with the best utilization of its properties. Ideally, the recyclables are directed to material recycling, the calorific fraction is directed to energy recovery, the biodegradables are directed to compost production. The modelling results demonstrated that this benefit cannot be realized without separate collection of food waste. Compost produced via any other collection system cannot be guaranteed to meet sanitary standards, irrespective of expenses for purifying, and cannot be sold. The only possible way to use it is covering waste at landfill sites. In ecological terms, this is equal to landfilling. In this situation the most environmentally soundable scenario comprises recyclables separation and subsequent WTE treatment of all the rest except for some non-combustibles.

4. Conclusion

Modelling MSW treatment scenarios for St. Petersburg and Leningradskaya oblast' demonstrated the following results. The "zero" scenario, relied on sanitary landfilling, shows the highest environmental impact in all the categories considered, thus, the scenario cannot be recommended. The "industrial processing" scenarios allow reducing this impact substantially. The complex scenarios, including both MBT and energy recovery, demonstrate the highest damage reduction. Meanwhile, the main benefit of complex treatment cannot be fully realized without the manufacture of marketable compost. Unfortunately, this cannot be done without a separate collection of food waste. In this situation the most environmentally soundable scenario comprises recyclables separation and subsequent WTE treatment of all the rest with the exception of some non-combustibles.

The proposed LCA based approach allows evaluating environmental efficiency, comparative weighting impact categories, defining efficient ways to minimize the impact for waste treatment methods and scenarios. The approach can be extended to other regions, which will improve effectiveness of decision-making while developing regional waste management systems.

Acknowledgments

The work has been carried out with financial support from the Russian Science Foundation, grant number 19-79-10114 (project "REC Mekhanobr-Tekhnika").

References

[1] On approval of the territorial system for the waste management The decree of the Government of St. Petersburg 193-r from 13.07.2020

[2] The territorial system for the waste management The decree of the Department of Leningrad Oblast for organization and control of the waste treatment system 5 from 22.07.2019

[3] Vaisberg L, Mikhailova N and Yasinskaya A 2019 Tendencies in the Development of Industrial MSW Processing in Western Europe *Ecology and Industry of Russia* **23(12)** 41–47

[4] Chusov A, Neguliaeva E and Romanov M 2018 Optimization of the Solid Waste Management System in Saint Petersburg Based on the Morphological Composition Study *MATEC Web of Conferences* **193** 8

[5] He P J, Pu H-X, Shao L-M and Zhang H 2017 Impact of co-landfill proportion of bottom ash and municipal solid waste composition on the leachate characteristics during the acidogenesis phase *Waste Management* **69** 232–241

[6] Moody C M and Townsend T G 2017 A comparison of landfill leachates based on waste composition *Waste Management* **63** 267–274

[7] Robinson H D, Knox K, Bone B D and Picken A 2005 Leachate quality from landfilled MBT waste *Waste Management* **25(4)** 383–391

[8] Molleda A 2020 Release of pollutants in MBT landfills: Laboratory versus field *Chemosphere* **249** 126–145

[9] Öman C B and Junestedt C 2008 Chemical characterization of landfill leachates – 400 parameter and compounds *Waste Management* **28** 1876–1891

[10] Yasinskaya A V 2021 Environmental impacts of disposal methods used for solid waste and its processing products *Obogashchenie rud* **1** 37–43 DOI: 10.17580/or.2021.01.07
[11] Vogtländer J G, Bijma A and Brezet H C 2002 Communicating the eco-efficiency of products and services by means of the eco-costs/value model *Journal of Cleaner Production* **10(1)** 57–67
[12] Indicator systeem ecocost. Delft University of Technology Available at: https://www.ecocostsvalue.com
[13] *Temporary method for determining prevented environmental damage* 1999 (Moscow)
[14] Oonk H 2012 Efficiency of landfill gas collection for methane emission reduction *Greenhouse Gas Measurement and Management* **2(2-3)** 129–145
[15] Zhenhan D, Scheutz C and Kjeldsen P 2020 Trace gas emissions from municipal solid waste landfills: A review *Waste Management* **119** 39–62
[16] Santos I F C, Barros R M and Filho G L T 2020 Biogas Production From Solid Waste Landfill *Encyclopedia of Renewable and Sustainable Materials* 11–19
[17] Fei X, Zekkos D and Raskin L 2016 Quantification of parameters influencing methane generation due to biodegradation of municipal solid waste in landfills and laboratory experiments *Waste Management* **5** 276–287
[18] Karapidakis E S, Tsave A A, Soupios P M and et al 2010 Energy efficiency and environmental impact of biogas utilization in landfills *International Journal of Environmental Science and Technology* **7** 599–608
[19] LFG Energy Project Development Handbook. EPA March 2020 Available at: https://www.epa.gov/sites/production/files/2016-11/documents/pdh_full.pdf
[20] Marcotte A and Fleming I R 2020 Damage to geomembrane liners from tire derived aggregate *Geotextiles and Geomembranes* **48(2)** 198–209
[21] R&D Technical Report P1-500/1/TR The likely medium to long-term generation of defects in geomembrane liners, January 2004, *Environment Agency* p 160
[22] Rotter V S, Kost T, Winkler J and Bilitewski B 2004 Material flow analysis of RDF-production processes *Waste Management* **24** 1005–1021
[23] Shumal M, Jahromi A, Ferdowsi A, Dehkordi S and et al 2020 Comprehensive analysis of municipal solid waste rejected fractions as a source of Refused Derived Fuel in developing countries (case study of Isfahan – Iran): Environmental Impact and sustainable development *Renewable Energy* **146** 404–413
[24] Viczek S A, Lorber K E, Pomberger R and Sarc R 2021 Production of contaminant-depleted solid recovered fuel from mixed commercial waste for co-processing in the cement industry *Fuel* **294** 120414
[25] Du Y-J, Shen S-L, Liu S-Y and Hayashi S 2009 Contaminant mitigating performance of Chinese standard municipal solid waste landfill liner systems *Geotextiles and Geomembranes* **27(3)** 232–239
[26] Giroud J P and Bonaparte R 1989 Leakage through Liners Constructed with Geomembrane – Part I. Geomembrane Liners *Geotextiles and Geomembranes* **8** 27–67

Method for milk whey microfiltration with filtrate pulsed backpressure and installation for its implementation

N N Korysheva[1], S A Titov[2], S V Shakhov[1], G S Tikhonov[3], I A Glotova[3], and N A Galochkina[4]

[1]Department of Machines and devices of food production, Voronezh State University of Engineering Technologies, 19, Revolution avenue, Voronezh, 394036, Russia
[2]Department of Physics, Heat Engineering and Heat Power Engineering, Voronezh State University of Engineering Technologies, 19, Revolution avenue , Voronezh, 394036, Russia
[3]Department of technology of storage and processing of agricultural products, Voronezh State Agrarian University named after Emperor Peter the Great, 1, Michurina st., Voronezh, 394087, Russia
[4]Department of Merchandising and Goods Expertise, Voronezh State Agrarian University named after Emperor Peter the Great, 1, Michurina st., Voronezh, 394087, Russia

E-mail: glotova-irina@yandex.ru

Abstract. During deep processing of whey using microfiltration, the loss of membrane efficiency can take place. In this work, an installation for microfiltration of milk whey has been developed. It includes pumps, containers with liquids, throttling valves, a pressure gauge, and a microfiltration cell with a tubular ceramic membrane. A thin titanium oxide layer was deposited on the inner surface of the porous alumina tube. The outer diameter of the tubes is 10 mm, the wall thickness is 2 mm, the length of the tubes is 45 cm. A homogenized aqueous dispersion of sugar beet fiber was used as an agent that improves the performance of the installation by creating a pulsed backpressure of the filtrate. It is shown that the use of a finely ground suspension of dietary fiber during microfiltration of milk whey through a tubular ceramic membrane prevents the formation of protein deposits on the membrane and in its pores. The installation allows obtaining a suspension of dietary fiber, enriched with milk protein, as an additional product. The protein-enriched fiber suspension left over after microfiltration can be used in food production, for example, as a thickening agent in the production of yogurts.

1. Introduction

The main issue of milk whey production is its potential to harm the environment. The whey potential for environmental contamination is 100-175 times greater than that of domestic wastewaters. In the world, the largest volumes of milk whey are obtained in Europe, North America, and Oceania, which is due to the developed production of cheese. At the same time, there is no growth in the production of dry whey, which is due to a much greater interest in baromembrane technologies and nanotechnologies as applied to the processing of this raw material. Similar trends are typical for many countries that are world leaders in milk production and dairy processing [1, 2].

The intensive processing methods (including the concentration of nutrients and dehydration) are used to store and transport milk whey. In addition, fractionation of useful substances of whey using membrane processing makes it possible to obtain products that are actively used in agriculture, food industry, and medicine (Fig. 1). These products include demineralized whey (DWP), whey protein concentrates (WPC), whey protein isolates (WPI), lactose, phosphates (PP).

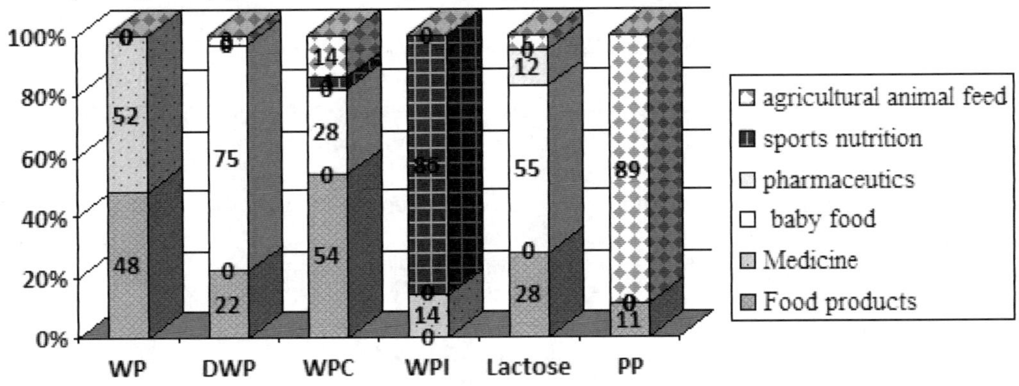

Figure 1. The structure of whey application and its fractionation products

The initial stage in the technology of deep processing of whey is microfiltration using semi-permeable membranes. Microfiltration of whey is often used to increase its transparency, reduce the number of microorganisms, and remove suspended protein particles. The microfiltrated whey can be used to make various drinks, jellies, and other food products. In addition, microfiltration is a preliminary stage for some types of ultrafiltration and nanofiltration of whey. However, the performance of microfiltration installations is limited by the deposition of protein on the membrane surface, as well as the formation of protein deposits on the pore walls. This is especially evident in the operation of ceramic membranes with a complex and strongly curved pore shape.

Predominantly, the aqueous lactose-salt fraction of whey passes through the membrane, since permeability and selectivity are the performance characteristics of the membrane. As a result, a boundary layer is formed at the membrane surface, which has an increased concentration of high molecular weight substances in comparison with the bulk of whey. This phenomenon harms the operation of the baromembrane treatment units of dairy systems and limits their use at the milk processing enterprises.

Electro-ultrafiltration [2], ultrasonic radiation [3], preliminary electroflotation treatment of milk whey [4] were previously used to reduce contamination of the membrane surface with protein deposits.

Nowadays, the combination of adsorption technologies with ultrafiltration is actively explored. To reduce contamination of the membrane surface, three processes are usually used. The first one is the preliminary adsorption process, in which a settler is required after adsorption, and the adsorbents are precipitated before the filtration system. The second one is the direct filtration process, in which raw water is fed into the membrane tank immediately after adsorption. The third is the pre-deposition process, in which adsorbent layers are pre-deposited on the membrane surface before filtration [5, 6].

The first two processes have already been studied and found some practical application. The third process was investigated only in a few works, and only in relation to flat membranes [7, 8]. This method of using adsorption effects in combination with ultrafiltration was used only for the purification of drinking water, for example, from impurities [9,10].

For the same purpose, studied was the influence of pre-precipitated flakes of aluminum salts on the contamination of ultrafiltration membranes with humic substances and proteins [11]. Compared to the processes of preliminary adsorption and direct filtration, the technology with preliminary deposition of the adsorbent on the membrane demonstrated much better specific removal of organic substances from the filtered solution [12].

Activated carbon is most often used as an adsorbent [13]. However, it can cause damage to the membrane due to the chipping of particles from its surface during operation for a sufficiently long time [14]. Tubular membranes are more convenient to use as compared with flat ones due to their durability, ease of washing and disinfection, and favorable hydrodynamic conditions in the channel of the ultrafiltration unit [15]. At the same time, for the filtration of food media, it is desirable to use some food material of the dynamic membrane so that both the filtrate and the concentrate can be eaten.

The work is based on the hypothesis that the creation of pulsed backpressure of the filtrate in the installation for microfiltration of milk whey with ceramic membranes can solve the problem of concentration polarization, which is described in detail in [1].

This work aims to develop and study a method for reducing concentration polarization during the processing of milk whey using microfiltration, as well as an installation for its implementation with the possibility of creating pulsed backpressure of the filtrate.

2. Materials and methods

Taking into account the tendencies of expanding the range of products based on goat milk, samples of cow and goat whey were used as objects of research. The characteristics of the objects of study are presented in Table 1. The physicochemical parameters of the milk whey samples were determined by standard methods.

Table 1. The physicochemical parameters of the milk whey samples

No. of sample	Characteristics	pH	Weight fraction of protein, %	Acidity, °T	Density, kg/m^3
1	Goat cheese wheat	6.25±0.02	1.16±1.0	14.0±1.0	1024.0±1.0
2	Goat curd whey	4.11±0.02	1.10±1.0	62.0±1.0	1029.0±1.0
3	Goat curd whey	4.33±0.02	0.80±1.0	61.0±1.0	1027.0±1.0
4	Cow cheese wheat	4.81±0.02	1.27±1.0	47.0±1.0	1026.0±1.0
5	Cow cheese wheat	5.81±0.02	0.98±1.0	16.0±1.0	1025.0±1.0
6	Goat curd whey	4.26±0.02	1.20±1.0	60.0±1.0	1026.0±1.0
7	Cow cheese wheat	6.26±0.02	1.32±1.0	14.0±1.0	1027.0±1.0
8	Goat cheese wheat	5.89±0.02	0.80±1.0	10.0±1.0	1020.0±1.0

This work studies the possibility of microfiltration of curd whey using the installation with pulsed backpressure of the filtrate (Figure 2). In this case, a homogenized suspension of beet fibers was used as an agent that improves filtration performance and, in particular, productivity.

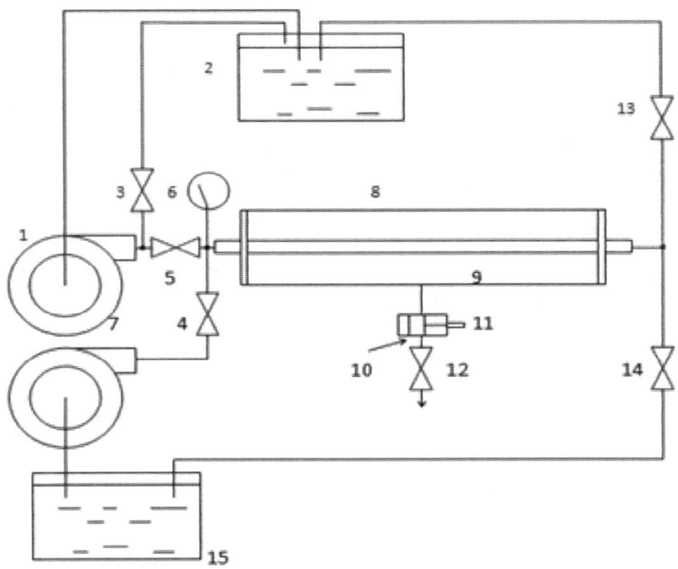

Figure 2. The installation for microfiltration using pulsed backpressure of the filtrate: 1, 7 - pumps; 2, 15 - containers with liquid; 3, 4, 5, 12, 13, 14 - throttling valves; 6 - manometer; 8 - microfiltration cell; 9 - tubular membrane; 10 - valve; 11 - valve piston.

The tubular ceramic membrane (9) was developed by Tupolev-Plaza. A thin titanium oxide layer was deposited on the inner surface of the porous alumina tube. The outer diameter of the tubes was 10 mm, the wall thickness was 2 mm, the length of the tubes was 45 cm.

Samples of ceramic membranes were examined using the JEOL-2SM-6380LV scanning electron microscope (Figure 3). A 7 mm long tube section was sawn along with a special high-speed disk cutter. A thin layer of gold was deposited onto the titanium surface under study by vacuum deposition to drain electrons.

Figure 3. Electron microphotograph of the membrane

3. Results and discussion

Initially, the experiments were carried out only with the fiber suspension of 0.5% concentration. As can be seen from Figure 4 (column 2), when this suspension is filtered, less permeate is formed than when pure water is passed through the membrane under the same conditions (column 1). So, during filtration, a polarizing layer is formed, consisting of fibers, and serving as a second membrane, which creates some additional obstacle for the water flow passing through the formed system of membranes. The system depressurization with an increase in the tangential flow rate ("flushing") and the subsequent pouring of pure water into the system instead of a fiber solution with further washing with water leads to an increase in the normal water flow during the subsequent control passage of pure water through the membrane (we conventionally call this operation "water filtration") (column 3). Consequently, the polarizing layer is partially washed away by the tangential flow of water in the absence of a through-flow pressing the fiber particles to the membrane. After the secondary formation of the polarization layer by filtration of the fiber suspension (column 4), a backflow was performed, as a result of which the filtration rate increased (column 5). Backflow loosens the polarizing layer and, in addition, washes away fiber particles that clog some of the membrane pores. Subsequent washing leads to an even greater increase in the filtration rate (column 6).

Comparison of fiber particle sizes (1-2 μm) determined by light microscopy and membrane pore sizes measured by scanning electron microscopy (0.1 μm) is shown in Figure 3. It confirms that the formation of a polarizing layer plays a decisive role in reducing the through-flow of liquid through the membrane, but not clogging of pores with fiber particles.

Further experiments included passing the whey through a layer of fibers previously formed as described above. To form this layer, the fiber suspension was filtered through a membrane. As in previous experiments, this led to a decrease in the filtration rate V_f (Fig. 5, column 2). During the subsequent filtration of whey, it was found that V_f decreases even more (column 3). This is due to the adhesion of the protein particles to the fibers in the polarizing layer. As a result, the steric hindrance for whey passage through the fiber layer is created.

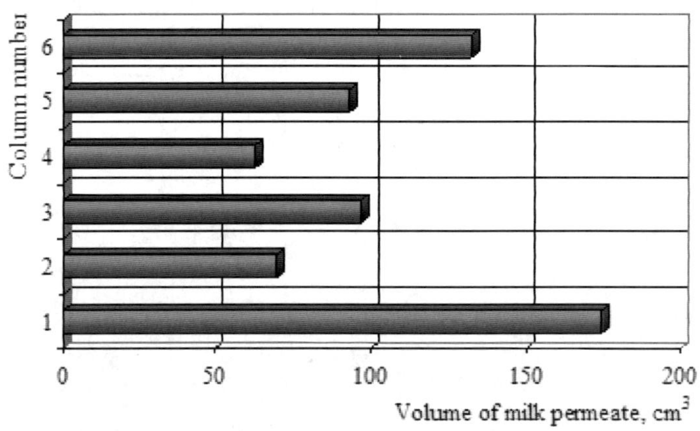

Figure 4. The volume of permeate per minute when filtering a suspension of dietary fiber in water under various conditions: 1- water, 2 - gel + fibers, 3 - water (washing), 4 - water (washing + purging), 5- water (washing), 6 - gel + fibers (layer formation).

Passing water through the formed filtration system (column 4) showed that the filtrate contains protein, that is, the protein is partially washed out from the fiber layer. As a result, the secondary "water filtration" was faster (column 5). The pulsed backflow further increased the flow of water through the filter (column 6). The presence of protein in the filtrate when water is passed through the membrane after whey filtration and its absence when whey is passed is apparently associated with the different acidity of water and whey. The isoelectric point of whey proteins is on average about 5.0, that is, in acidic whey (pH 4.5), the protein molecules are positively charged and are attracted to negatively charged fibers. In water (pH 7), proteins become negatively charged, repel from the fibers, and enter the permeate.

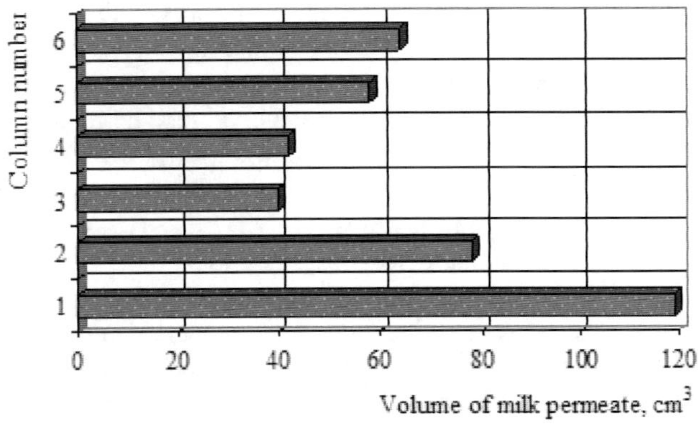

Figure 5. The volume of permeate per minute when filtering whey through the polarizing layer of dietary fibers: 1 - water, 2 - gel, 3 - whey, 4 - water, 5 - washing (water with protein residue), 6 - water.

The next stage of experiments examined the filtration of a suspension of dietary fiber in whey (Fig. 6). During this filtration, a polarizing layer is also deposited on the membrane, resulting in a reduced filtration rate compared to "water filtration" through a clean membrane (column 1). As in the previous experiment, "water filtration" through the resulting system (column 2) leads to partial washout of protein and an increase in V_f when water is subsequently passed through the filter (column 3), after which the pulsed backflow increases the filtration rate (column 4). Further washing of the system with water leads to an additional increase in V_f (column 5). During the subsequent filtration of the suspension of dietary fiber in the whey, another layer is overlapped on the partially destroyed and washed off layer, so the filtration rate decreases (column 6). Further treatment of membranes with a pulsed backflow (column 7) and washing (column 8) partially destroys the secondary layer.

If after filtration of the suspension there is no "filtration of water", then the protein molecules in the layer of fibers strengthen this layer, which connects the individual fibers, therefore, as individual experiments have shown, the pulsed backflow does not destroy this layer. However, flushing with water has some effect, the filtration rate is increased by about 40%., The fiber content does not significantly affect the filtration rate for fiber concentrations of 50-250 g/l, therefore it is beneficial to perform filtration at a minimum concentration of 50 g per liter of whey.

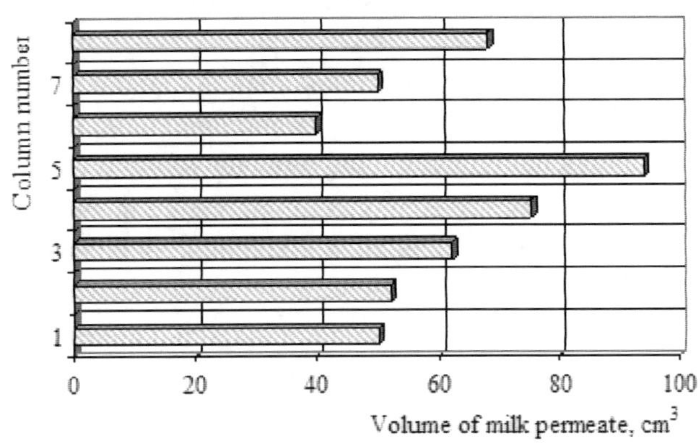

Figure 6. The volume of permeate per minute when filtering a suspension of fibers with milk whey: 1 - whey + fibers; 2 - fibers; 3 - fibers (second run); 4 - purging; 5 - flushing; 6 - whey + fiber; 7 - fibers; 8 - fibers.

When this process was performed in a cyclic mode (one cycle consists of filtration of the whey-fiber mixture, purging, and flushing with water and lasts 7 minutes), the maximum filtration rate in the cycle (about 60 cm^3/min) was approximately constant and did not depend on the number of cycles (Fig. 7). This is because the fiber layer prevents protein from entering the pores of the membrane.

Spectral analysis of the filtrate was performed after each cycle. The measurements were carried out in the wavelength range 220-310 nm. The obtained UV absorption spectra of the permeate collected after the second and third filtration cycles (Fig. 9) had a peak at λ = 275 nm, which is due to the absorption of tryptophan groups of protein molecules.

The spectrum of the first cycle permeate had no peak at λ = 275 nm, so, this liquid contains practically no protein (Fig. 8). This was due to the fact that after the first cycle, the fiber layer was saturated with protein, that is, the protein molecules were deposited on all adsorption centers of the layer, while whey that had passed through the fiber layer was cleaned from protein. The subsequent passing of whey through the fibers no longer leads to protein binding, since all adsorption centers were already occupied.

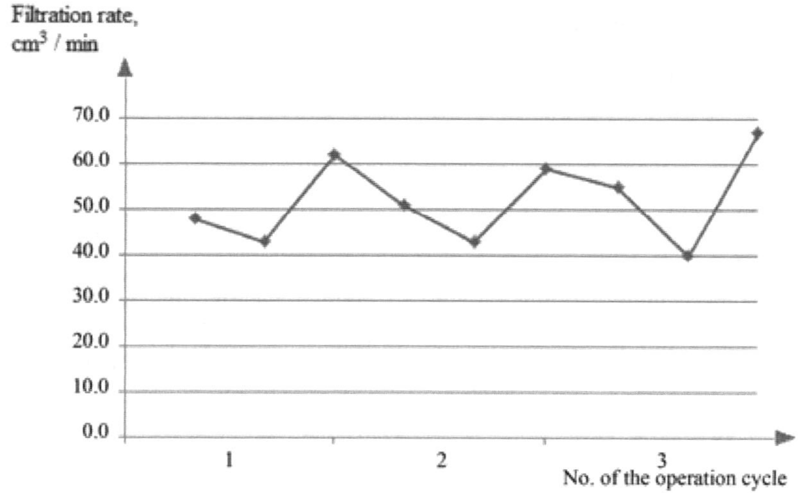

Figure 7. Change in filtration rate (ml/min) during cyclic operation of the installation.

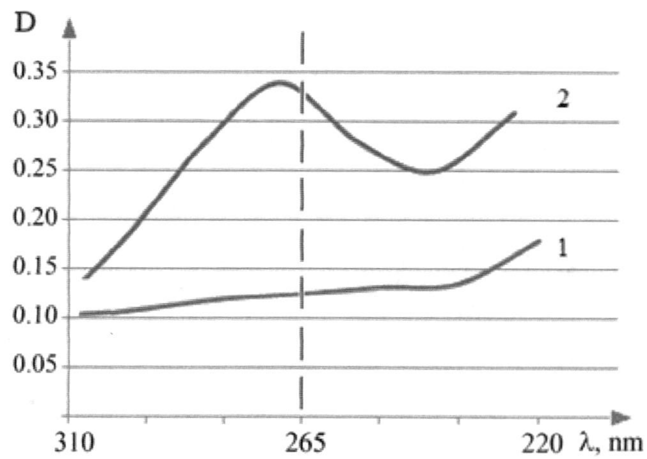

Figure 8. Absorption spectra of permeate after filtration cycles: 1 - first; 2 - second and third.

4. Conclusions

Thus, the studies have confirmed the feasibility of the techniques described in the work. The proposed design of a pilot installation using pulsed backpressure of the filtrate created by a homogenized aqueous suspension of dietary fiber can increase the efficiency of whey microfiltration and can be recommended for use in enterprises with different capacities. The protein-enriched suspension of dietary fiber remaining after microfiltration can be used in food production, for example, as a thickening agent in the production of yogurts [16].

References

[1] Meyer P, Hartinger M, Sigler S and Kulozik U 2017 *Food and Bioprocess Technology* **10** 674–686
[2] Lazarev S I, Kovalev S V, Kovaleva O A, Rodionov D A, Lazarev D S and Konovalov D N 2019 *Chemical and Petroleum Engineering* **55(1-2)** 114–121
[3] Tejaswini P, Rajeshree A K and Parag R 2019 *Ultrasonics Sonochemistry* **55** 125–134
[4] Evdokimov I A, Titov S A, Polyanskyd K K and Saiko D S 2010 *Foods and Raw Materials* **5(1)** 131–136
[5] Malczewska B, Liu J and Benjamin M M 2015 *J. Membr. Sci.* **479** 159–164
[6] Cai Z X and Benjamin M M 2011 *Environ Sci. Technol.* **45(20)** 8935–8940
[7] Thiruvenkatachari R, Shim W G, Lee J W and Aim R B 2006 *Colloids and surfaces A: Physicochem. and Eng. Aspects* **274(1–3)** 24–33
[8] Ajmani G S, Goodwin D, Marsh K, Fairbrother D H, Schwab K J, Jacangelo J G and Huang H O 2012 *Water Research* **46(17)** 5645–5654
[9] Ran Zh, Yao M, He W and Wang G 2020 *Separation and Purification Technology* **249** 117115
[10] Du X, Qu F and et al 2014 *Chem Eng. J.* **254** 293–301
[11] Ma B and Wang X 2017 *J. Membr. Sci.* **538** 34–40
[12] Kim J, Cai Z X and Benjamin M M 2010 *J. Membr. Sci.* **349(1–2)** 35–43
[13] Tarasevich Yu I 1999 *Studies in Surface Science and Catalysis* **120** B 659–722
[14] Cai Z X, Wee C and Benjamin M M 2013 *J. Membr. Sci.* **433** 32–38
[15] Lazarev S I, Abonosimov O A, Rodionov D A, Ignatov N N, Levin A A and Bogomolov V Yu 2018 *Chemical and Petroleum Engineering* **54** 405–409
[16] Ponomarev A N, Melnikova E I, Bogdanova E V and Kharitonov D V 2017 *Foods and Raw Materials* **5** 41–50

Combination of ground penetrating radar and seismic surveys of the tailings dam

A Kalashnik[1] and A Dyakov[1]

[1]Mining Institute Kola Science Centre RAS, Fersman st. 24, Apatity, Russia

E-mail: a.kalashnik@ksc.ru

Abstract. Identification of water-saturated zones in the tailings dams is an actual scientific and practical task in terms of providing, first of all, their mechanical strength and filtration stability. Use of active sounding geophysical study methods allows obtaining sufficiently detailed information about the peculiarities of the internal structure of the tailings dam and the degree of water saturation of the composing soils. The paper presents the results of in-situ experiments on the study of the tailings dam of the mining enterprise by ground penetrating radar (GPR) and seismic methods. A comparative analysis of the conducted studies has allowed clarifying the internal structure and assessing the dam's condition, paying special attention to the identification of local zones of increased water saturation and filtration. Based on the calculated correlation coefficient of electromagnetic and seismic wave velocity values, it was revealed that synchronization of geophysical surveys allows significantly increasing the reliability of in-situ determinations, as well as obtaining more reliable data. The results of the studies are the basis for predicting the most vulnerable places (zones) of a bulk ground hydraulic facility, as well as the localization of water-saturated areas in the body of the ground structures with greater reliability and performance.

1. Introduction

Currently, the main trends of geophysics in the study of technogenically disturbed, including partially water-saturated, rocks and soils is to increase the information content and reliability of the data obtained [1-4]. A large number of applied geophysical methods indicate the absence of a single standard geophysical method, capable of providing prompt and high-quality obtaining the required information about the mining-geological environment, due to the complexity and ambiguity of the data obtained. At the same time, the joint use of different in nature wave methods: electromagnetic (GPR) and seismic profiling allows receiving more reliable operative information about internal structure and degree of water saturation of technogenically disturbed rocks and soils [1, 5, 6]. Therefore, the choice and integration of these geophysical methods can improve the performance and reliability of the data obtained in solving various problems [7-9].

In mining operations, a large number of industrial facilities are potentially hazardous and are therefore classified as hazardous and technically complex facilities that require regular inspections and monitoring. One of such industrial facilities is the tailing dumps with systems of embankment structures, a breach of which stability and functionality can lead to an abnormal (or emergency) situation in the technological mining chain.

The world practice of mining nature management has typical examples of accidents at the tailing dumps, which led to significant financial and socio-economic consequences [7, 10]. A comprehensive examination and monitoring of the internal structure and condition of the tailings with the use of operational geophysical methods creates an information and technical basis for minimizing the risks of man-made accidents. The purpose of the paper is to substantiate methodological approaches to the combination GPR and seismic surveys of the tailings dam.

2. Methods

To assess the state of water saturation and the internal structure of the technogenically disturbed rocks of the tailings dam, two cycles of integrated observations were carried out using seismic and GPR profiling methods [11-13]. The following indicators were taken for further analysis: for the seismic method - the ratio of longitudinal velocity to transverse velocity *Vs*/*V*p; for the GPR method - electromagnetic wave velocity *V*.

3. Results

Figure 1 shows the main results of the performed integrated research as a radarogram of the electromagnetic wave velocity *V* (a) and a seismic tomogram of the ratio of longitudinal to transverse velocity *Vs*/*V*p (b). The Figure visually notes the similarity in the colour palettes reflecting the internal structural features and water saturation of the surveyed soil section. For quantitative comparison of seismic and georadar values, plots of variability of the accepted controlled parameters of the obtained results along the profiling traces in separate sections at two pickets (PK) (figure 2, 3) were built.

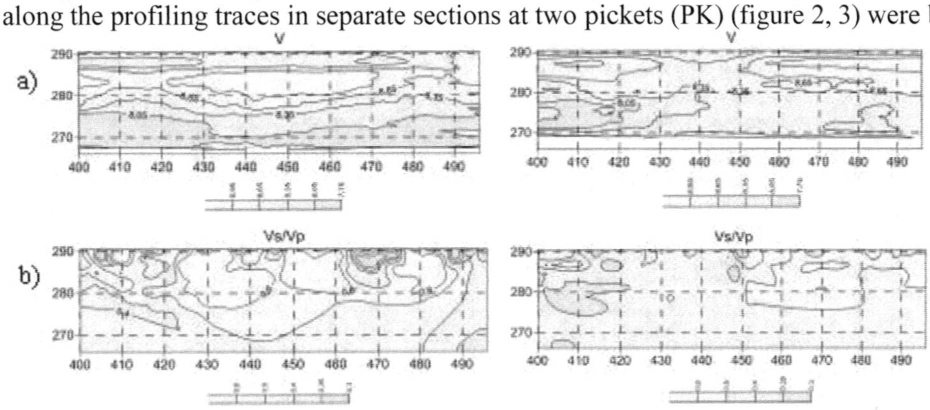

Figure 1. Results of integrated studies of soils with GPR (a) and Seismic (b) profiling methods (left-1st study cycle, right-2nd study cycle).

4. Discussion

Interpretation of the results of the first cycle of combined seismic and GPR measurements (figure 1, left) suggests the following: In the central part, both the radarogram and the seismogram, the distribution of data reflecting the presence of dense soils of natural moisture has a fairly good visual convergence.

At the same time, in some areas, the colour palettes of the radarogram and seismogram reflect a visual discrepancy in the nature of the distributions of the values of the adopted indicators: *V* and *Vs*/*V*p. On the radarogram we can clearly stand out the near-surface zone up to 2 m thick on two intervals of the profile (400-430 m and 470-480 m), in which the wetted soils prevail. At the same time on the seismogram the near-surface zone of moist soils is localized by small areas up to 5 m, and only at the end of the profile, at the interval of 470 m, a direct coincidence with the georadar profiling data is observed. In turn, according to seismic profiling data, water-saturated grounds are identified at the very end of the profile, in the interval 485-498 m, distributing over the whole depth of the seismogram, and according to GPR data, water-saturated grounds are identified throughout the profile, from a depth of 15 m and more (absolute marks 270-275 m).

These discrepancies can be explained, first of all, by the fact that seismic and georadar profiling were not performed simultaneously, but with a time difference of several days [14, 15]. During this period, the influence of natural (abundant precipitation such as rain) and technogenic (discharge of industrial water from an operating enterprise) factors, which also cannot be excluded, could have had an effect.

Interpretation of the analysed results of the 2nd cycle (figure 1, right), with synchronous seismic and GPR measurements, suggests the following. On the radarogram, the near-surface zone (to a depth of 1-1.5 m) of the profile up to 289 m, exposed to surface sediment penetration, is characterized by initial low velocity V = 8.1-8.22 cm/ns, with its further sharp increase to V = 8.36-8.71 cm/ns at 8-10 m depth from surface (marks 282.5-280 m). The most likely, this is caused by density growth and decrease of ground humidity.

In the interval of heights 280-270 m there is a zone of sharp fall of velocity values to the values V = 7.35-7.71 cm/ns. Similar changes of Vs/Vp index are noted on the seismogram at the same marks, the values of which are 0.30-0.35, which allows us to conclude that soils here are subjected to considerable water saturation.

To compare the values of seismic and georadar results, we plotted the correlation between the values of the monitored parameters in the same sections at two pickets (PK) (figure 2, 3), for the first and second cycles of measurements, respectively.

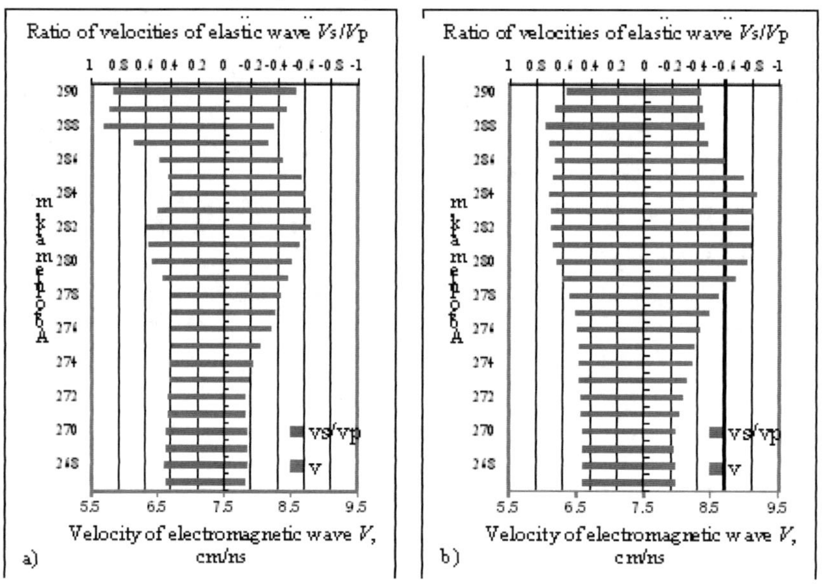

Figure 2. Comparison of seismic and georadar results of the 1-st measurement cycle: a) PK4+10m, b) PK4+60 m.

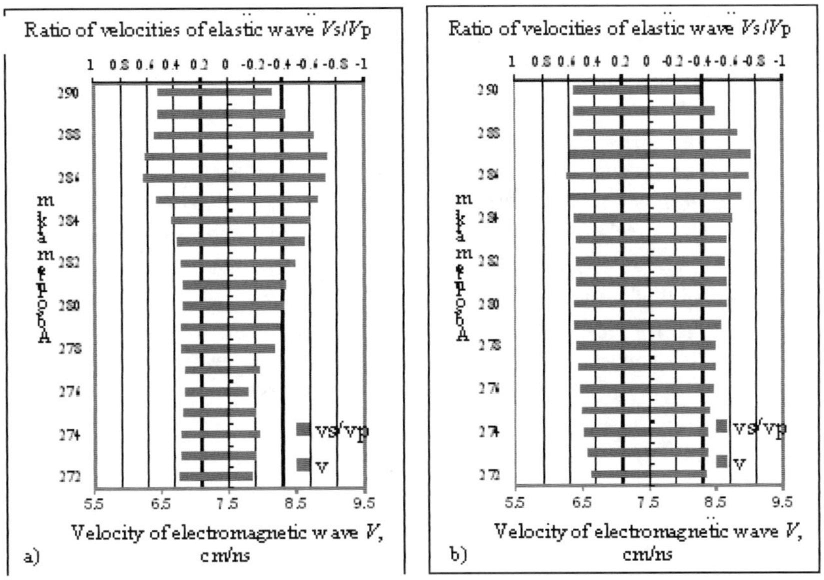

Figure 3. Comparison of seismic and georadar results of the 2-nd measurement cycle: a) PK4+10m, b) PK4+76 m.

Comparison of georadar and seismic profiling data, both on the first cycle and on the second, as well as in the aggregate on two cycles of measurements, showed the presence of confident correlation (convergence) of the obtained results. At the same time, GPR and seismic data of the second cycle of measurements have closer correlation (figure 3), which is caused by synchronization of measurements by two methods, in spatial-temporal coordination. In general, the correlation ratio calculated for the two measurement cycles has a range of values from 0.17 to 0.83 (except for individual values), with a predominant number of values from 0.4 to 0.6 or more (figure 4).

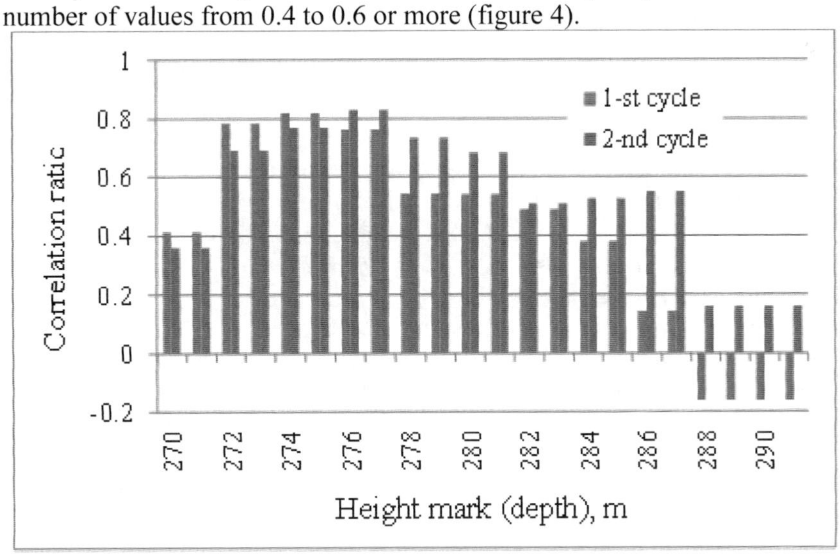

Figure 4. Correlation ratio of studied indices of GPR and seismic surveys.

5. Conclusions

To study the internal structure and state of technogenically disturbed soils of the tailings dam we carried out their research with a combination of GPR and seismic profiling, in two cycles: with a difference of several days and synchronously. Results of executed integrated surveys have allowed getting more reliable information about internal heterogeneity of technogenically disturbed soils due to unevenness of their water saturation. Zonal filtration heterogeneity of soils was revealed and their structure was clarified in the investigated area.

A comparative analysis of the values of the adopted indicators was performed: for the seismic method - the ratio of longitudinal velocity to transverse velocity Vs/Vp; for georadar method - electromagnetic wave velocity V. Good convergence of analysed indices for both cycles was revealed. At that, on the basis of the calculated correlation ratio of the analysed values, it was discovered that the synchronization of GPR and seismic surveys allows significantly increasing the reliability of in-situ studies, as well as to obtain more reliable data. The obtained results and new knowledge serve an information-technical basis for integrating of georadar and seismic profiling of technogenically disturbed soils and substantiation of methodical approaches to comparative analysis of their results.

References

[1] Starovoitov A V 2008 *Ground penetrating radar data interpretation* (Moscow: MMU Publ.) p 192

[2] Maruddani B and Sandi E December 2019 The Development of Ground Penetrating Radar (GPR) Data Processing *Intern. J. Machine Learning and Computing* **9(6)** 768–773

[3] Kumlu D and Erer I 2018 Clutter reduction in GPR images using non-negative matrix factorization *J. Electromagn. Waves and App.* **32(16)** 2055–2066

[4] Melnikov N N, Kalashnik A I, Zaporozhets D V, Dyakov A Yu and Maksimov D A 2016 Experience of applying GPR subsurface studies in the western part of the Russian Arctic sector *Problems of the Arctic and Antarctic* **1** 39–49

[5] Kalashnik A I and Zaporozhets D V 2020 Georadiolocation studies at mining enterprises of Kola region *Conference Proceedings Engineering and Mining Geophysics 2020 EAGE* 51089

[6] Demetrius Cunha Gonçalves da Rocha, Marco Antonio da Silva Braga and Camilla Tavares Rodrigues 2019 Geophysical methods for BR tailings dam research and monitoringin the mineral complex of tapira – minas gerais *Revista Brasileira de Geofísica* **37(3)** 275–289

[7] Melnikov N N, Kalashnik A I, Kalashnik N A and Zaporozhets D V 2017 Application of modern methods for complex research of the state of hydraulic structures of the Barents Sea region *Proceedings of the Murmansk State Technical University* **20(1)** 13–20

[8] Del Sole L, Calafato A and Antonellini M 2020 Combining Ground-Penetrating Radar profiles with geomechanical and petrophysical in situ measurements to characterize sub-seismic resolution structural and diagenetic heterogeneities in porous sandstones *Marine and Petroleum Geology* (Northern Apennines, Italy)

[9] Comisi F, De Giorgi L and Leucci G 22–24 October 2020 Integrated use of GPR and TDR for wood permittivity evaluation *IMEKO TC-4 International Conference on Metrology for Archaeology and Cultural HeritageTrento* 1–4

[10] Jorge Luís Porsani, Felipe Augusto Nascimento de Jesus and Marcelo Cesar Stangari 2019 GPR Survey on an Iron Mining Area after the Collapse of the Tailings Dam I at the Córrego do Feijão Mine in Brumadinho-MG *Remote Sensing* **11(7)** 860

[11] Abramov N N, Dyakov A Y and Kalashnik A I 2019 Identification of Water-Saturated Zones in a Protective Hydraulic Earthen Structure by Synchronous Electromagnetic and Seismic Sounding *Power Technology and Engineering* **53(2)** 167–171

[12] Danilkin A A, Kalashnik A I, Zaporozhets D V and Maksimov D A 2014 Monitoring the condition of the protective dam in the mining zone of the secondary deposit of Kovdorsky GOK *Mining informational and analytical bulletin* **7** 344–351

[13] Kalashnik A I, Zaporozhets D V and Kalashnik N A 2013 Identification of filtration and deformation processes in the tailings dam *Herald of the Kola Science Centre RAS* **2** 13–17

[14] Kalashnik A I and Dyakov A Y 2017 Information technologies in problems of monitoring of hydraulic facilities of mining enterprises through subsurface GPR sounding *Mining*

informational and analytical bulletin **23** 283–291
[15] Chromcak J, Grinc M, Panisova J, Vajda P and Kubova A 2016 Validation of sensitivity and reliability of GPR and microgravity detection of underground cavities in complex urban settings: Test case of a cellar *Contributions to Geophysics and Geodesy* **46(1)** 13–32

An environmental engineering approach to wastewater treatment with bivalves in the coastal zone

AV Gudimov[1]

[1]Murmansk Marine Biological Institute (MMBI RAS), Vladimirskaya 17, Murmansk, 183010 Russia

E-mail: alexgud@mail.ru

Abstract. High tolerance to pollution and filtration capacity of some seston-eating bivalves and mussels of g. *Mytilus* first of all, principally determine mussel populations as a powerful natural biofilter. Both marine and freshwater mussels consume relatively large amounts of seston, feeding directly on the primary producers. Particles with contaminants extracted from ambient water during the filtration are accumulated in mussel pseudofaeces and faeces discharged after digestion. These specific particles of faeces and soft pseudofeaces become very suitable substrata for bacteria development with following contaminant destruction. Moreover, such biodeposits appeared to be the nutritious food for bottom detritophages, in particular of polychaetes and amphipods. Some fluid metabolites excreted by mussels, i.e. dissolved organic matter (DOM) have a pronounced biological activity to producers. In general, mussels possess high resistance to acute and chronic pollution stress, although their growth may slow down and even stops. In clean conditions food supply is the most important factor in determining activity and growth rate of mussels for northern environment. Mussel and some other aquaculture, as a relatively new application of environmental engineering is able to improve the recycling of nutrients from wastewater to bioproduction and greatly reduce the risk of environmental degradation in the coastal zone.

1. Introduction

Mussels are very common bivalve species for coastal waters of numerous countries. Mussels of the genus *Mytilus* are almost cosmopolitan due to very wide range of environmental conditions within they can survive. *Mytilus edulis* itself is a widely distributed and variable species in the northern hemisphere. In the Barents Sea blue mussel occurs under extreme sub-arctic conditions, where the winter temperature of the coastal water can be at or below 0°C.

Blue mussel as a sessile mollusc, which in majority occurs intertidally or in shallow sublittoral waters, is an ideal species for studying biological effects of pollution in the coastal zone of northern seas. Mussel populations are often dominated in coastal waters, particularly in the estuarine areas.

Occurring in both open seawater and brackish estuaries, especially where there is a significant water movement coupled with food influx, mussels live on a variety of substrata, such as rock, stones, dead shells and even any coarse sand or compacted mud, i.e. anywhere that provides a secure anchorage. Mussels are found attached to ships, pier pilings and harbour walls.

As a good food product mussels and some other intertidal bivalves can be collected readily. Bivalve mollusks such as oysters and mussels have been cultivated for centuries both in coastal waters of European and Asian countries. It is important to note that production processes in such seston-eating bivalves are accompanied by efficient (up to 98%) cleaning of the ambient water from suspended particles due to filtration activity of these bivalves.

In addition, mussel ability as a biosensor of the environment and a sentinel organism has been developed recently to online pollution control of the coastal waters and aquaculture [1-3].

Since mussel populations are associated with the coastal zone and often more abundant there than other bottom invertebrates, they eventually became the protective interface between civilized society and the sea. A relatively new environmental engineering approach to mussel aquaculture is suggested to make a powerful protection zone based on mussel biofilter along the shoreline of some sites (cities, harbors) of wastewater discharge or/and other anthropogenic load to coastal ecosystems. To apply the natural self-purification processes in mussel populations for treatment of wastewater and pollution removal as an advanced partnership with nature, the evaluation of mussel abilities in terms of physiological processes and 'toxic responses' demands certain attention. Some original and analytic results of the research on this new approach are presented in the paper.

2. Materials and methods

Ecophysiological research of mussels *Mytilus edulis*, *M. galloprovincialis* and *M. trossulus* were carried out on the Barents, Black and Baltic Seas respectively, for several years from 1985 to 2016 and continued partly up to the present time [see references in 1, 2, 3]. The most detailed experimental research was performed on the Barents Sea with blue mussel M. edulis. Some results of these experiments on mussel growth (unpublished earlier) and filtration (re-calculated our data by [1]) are presented here as strongly related to the general concept (illustrated originally in figure 1-2) of this analytical investigation. Important arguments obtained from the review of reliable sources are maintained but not covered the approach proposed.

3. Results and discussion

Key ecophysiological features and abilities of filter-feeding bivalves and marine mussels, especially, that can be applied to use their coastal aquaculture as a powerful biofilter to establish the protection zones, have been comprised to following arguments:

3.1. Filtration

The process of water filtration and filtration capacity of bivalves are the principal issue for the biofilter and must be clearly understood.

In the most frequently used an indirect method for measurements of the filtration rate which is defined as that volume of water completely cleared of particles in unit time [4, 5]. Mussels as most lamellibranchs obtain their food from the particulate matter, micro-organisms included microalgae, protozoa and bacteria or fine dispersed detritus, suspended in the water. They transport water through the gills, which serve as filters and retain the suspended particles.

Water enters the pallial cavity through the inhalant siphon before passing through the gill ostia and is expelled through the exhalant siphon, which is narrower than the inhalant and lies at the posterior edge of the mantle. By means of cilia the material held back on the gill surfaces is to be transported towards the labial palps. The palps regulate the amount of food which enters the mouth, and direct surplus material on to the rejection of the mantle surface. From the palps, food is directed into the mouth and ingested [6, 7].

3.2. Particles retention and food concentration

Bivalve filter- feeders, such as mussels, process large volumes of water and efficiently retain small suspended particles [8, 9]. At relatively low particle concentrations in the water the gills transport water at constant high rates and efficiently retain suspended particles of size down to a few μm [6, 7]. Cirri of the bivalve gill act as filters that retain suspended particles in the through current.

There are some evidences for complete or near-complete retention of particles larger than 2 μm by mussel gills, because the retention of fine particles by the meshwork of cilia of the latero-frontal cirri is accomplished by producing the mucus sheet, especially under conditions of poor food supply [9]. Vahl [10] has demonstrated almost 100 % retention of small motile flagellates of 1-2 μm diameters by undisturbed *M. edulis* in a flow-through seawater system.

A comparison between filtration intensity of the most active freshwater filter-feeders and marine bivalves has shown the highest values of filtration rate in mussels (table 1). This is evidently the result of adaptation to lower concentrations of suspended matter in the sea water.

Table 1. Filtration intensity of bivalves per g of unit wet (dry) weight tissues

Species	Filtration intensity, ml/h	Reference
Dreissena and Unio	85	[12]
Mytilus edulis	386	[8]
Mytilus edulis	(3900)	[10]
Mytilus edulis	(3200)	[1]
Arctica islandica	180	[4]
Modiolus modiolus	240	[4]

The maximum filtration rate is observed at very low suspension concentrations quantified normally in the natural environment [7, 11]. At the same time, maximal filtration rate in natural sea water is fluctuated from 2 to 4 l per h for a mussel of 4-5 cm in length [9].

Important to note, that increasing food concentration results in a pronounced reduction in filtration rate [5, 7, 9, 11]. In contrast to the assimilation rate this reduction is practically independent of the increased food concentration then, and to the amount of food consumed (ingestion rate). Edible part of ingested particulate matter is digested with following discharge of faeces and release of fluid excreta.

3.3. Biodeposition

At high food concentration a part of the algal cells is accumulated to pseudofaeces, and the filtration rate is reduced at once to a remarkable degree [9, 7]. As soon as concentration of the suspension twice exceeds the natural one the filtration activity declines and pseudofaeces appear [7, 12]. Their amount increases with increase of the suspension concentration, and some of the algae filtered out of suspension are rejected. Substances agglutinated in pseudofaeces and faeces, i.e. biodeposits, appeared to be the most suitable substrata for bacteria and then nutritious food for detritophages, in particular of polychaetes, amphipods and some other bottom invertebrates [7, 8, 12, 13]. Excreta of living molluscs are not only included into the detritus food web, but also influence the formation of primary production.

3.4. Stimulating of primary production

Metabolites excreted by mussels, i.e. dissolved organic matter (DOM) possess a certain biological activity in relation to producers. Thus, in regions of mussel cultivation and above their natural populations the numbers of bacterioplankton cells are 4-6 times and of phytoplankton cells 10-30 times as higher than in control areas. Experimental studies showed that blue mussels excreted about 12 % of their diet as DOM [14]. Excretion in mussels cultured on rafts was almost twice as much higher than that in natural populations of the Barents Sea.

Evidently DOM excreted by mussels influence phytoplankton in the native state, but not as nutrients, since the assimilation activity of phytoplankton cells has been increased to a considerable extent whiles the concentration of chlorophyll «a» practically has not. The excretion varies with the season and has the highest values in spring at a deficit of nutrients in the coastal water [14]. Thus, through the phytoplankton production mussels promote the recycle some of nutrients for the benefit of the ecosystem (figure 1).

On the other hand, it was concluded that an increase in the biomass of cultured mussels could lead to a decrease in primary production [15] and thus the consequences of eutrophication in the surrounding waters.

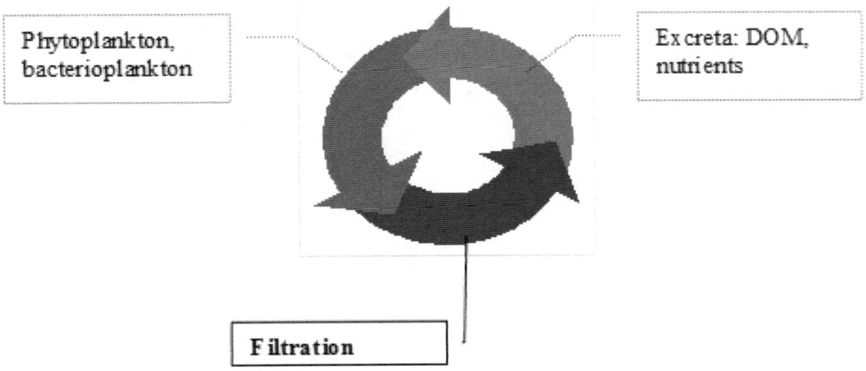

Figure 1. Recycling loop in a mussel-dependent ecosystem

3.5. Recycling

Mussels consume relatively large amounts of seston, feeding directly on the primary producers, phytoplankton and bacteria, and non-living organic material. In the summer and autumn (during 5-6 months in temperate climate) 20-30 % of filtered organic matter can be fixed as meat weight by M. edulis [9]. The net effect of mussel aquaculture will always be a reduction of nutrients in the area (with obviously strong mitigation or removal of eutrophication) despite of high release of phosphate, nitrate, ammonium and silicate from mussel bed into the water column observed [16]. It was proved that excretion by mussels contributed 31-85 % to the total phosphate flux from the mussel bed. About 75 % of the particulate organic P, taken up by the mussel bed, was recycled as phosphate. A calculation of total N remineralization showed that mussels play a major role in the recycling of nitrogen in the estuary [16]. Mineralization of biodeposits on the mussel bed is the main probable source of the regenerated nutrients.

3.6. Growth

Food supply is obviously the most important factor in determining activity and growth rate of mussels for northern environment [17]. It is thus suggested that in the Barents Sea the sub-optimal conditions for mussel growth take place even in summer time and variations in the growth rates occur according to local food conditions.

During the long, cold and dark winter concentration of the food valuable phytoplankton decreases drastically until it's almost full absence. The quality of seston as food for mussels changes for lesser content of proteins, while organic detritus supplies a large part of mussel's diet. At the low food concentration the metabolic reserves in mussels are utilized until more favorable conditions arise [9, 17]. The growth rate changes accordingly the food supply. Even mussels grown under apparently identical conditions can exhibit the widely different rates. The results obtained by the author (table 2) revealed that in the floating suspended mesh-cages the growth rate of adult mussels was from 2 to 5 times as higher than in the bottom layer of the upper sublittoral because of the limited food conditions on the arctic and sub-arctic coast.

Table 2. The monthly-averaged linear increments in mussel lengths (mm per month) after 12 months of growth in flow-through seawater aquaria with natural seston inflow from the subtidal zone and in the floating suspended cages located in the Dalnaya Zelenetskaya Bay, the Barents Sea

Subtidal conditions	Size classes, mm				
	10-20	20-30	30-40	40-50	50-60
Subsurface layer (cages)	1.2	1.0	0.8	-	0.56
Bottom layer (group #1)	0.7	0.5	0.13	-	0.14
Bottom layer (group #2)	0.74	0.55	0.32	0.13	0.11

The influx of nutrients from wastewater discharged to the coastal zone is able to increase significantly the primary production and evidently, the food supply for mussel growth.

3.7. Pollution removal with coastal biofilter

Because of their coastal and estuarine distribution, some wild mussel populations along with commercially exploited mussel beds and aquaculture can be subjected to sewage pollution.

The limitations of recycling in wastewater treatment are affected by the problems of toxic pollution effects. The untreated mixed wastewater is not suitable for environmental engineering with aquaculture. At the same time domestic and industrial sewage are often mixed into the water bodies. As a consequence of this the nutrient recycling concept stresses detoxification as a first stage of wastewater treatment, to avoid accumulation of, for example, metals in the food chain.

Mussel is of interest to ecologist studying pollution, as it comprises a sedentary filter-feeding invertebrate, which are readily to accumulate pollutants from the environment.. The lack of capacity to regulate internal concentration of elements leads to accumulation of heavy metals and other different anthropogenic chemicals in bivalve tissues to a marked degree over sea water levels. At the same time, mussels are tolerant to wide range of environmental contaminants [1, 9]. Owing to functioning of detoxification system [13] molluscs retain vital activity, although their growth slows down and even stops.

Even crude oil does not appear to be highly toxic to mussels. Blue mussels are able to concentrate and deposit oil residues as well as other contaminants, extracted from ambient water. In the process of filtration mussels accumulate contaminants [9] and oil [18] in faeces and pseudofaeces discharged after digestion.

Mussels which have been exposed to chronic sublethal concentrations of hydrocarbons accumulate these in their lipid pools. Lee et al. [19] suggested that the digestive gland was the site of hydrocarbon storage; they also found that on transfer to clean seawater, after previous exposure to petroleum hydrocarbons solutions, 80-90% of the hydrocarbons were eluted. The polycyclic aromatic hydrocarbons were initially released rapidly from the mussels then slower phase followed [20].

In fact, mussels should have the beneficial effect of rapidly clearing seawater and reducing contamination and even pathogens level in the coastal environment. Mussel tolerance to pollution allows considering a possibility of using them in an integrated biological purification as a coastal biofilter (figure 2) for removal of the suspended matter, nutrients, bacteria and viruses, petroleum hydrocarbons, heavy metals and other anthropogenic chemicals in the process of wastewater treatment, while protection of the environment is coupled with bioproduction in environmental aquaculture.

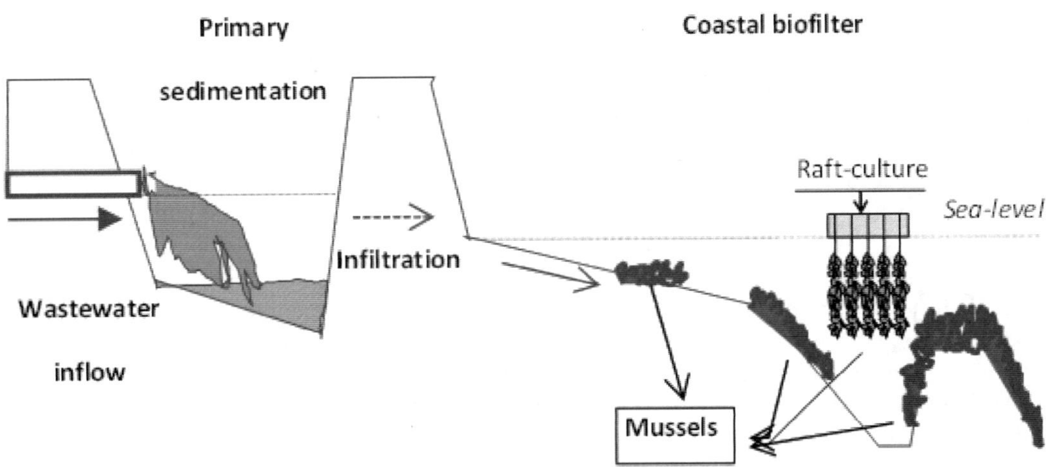

Figure 2. A principal sketch of wasterwater treatment by mussel biofilter consisted of natural littoral and sub-littoral populations, and mussel aquaculture. Primary sedimentation/infiltration is an important stage of the wastewater treatment to avoid the negative effect of suspended matter surplus on mussel activity and growth

4. Conclusions

In applied ecology mussels are well-known as a reliable biological indicator and monitor of pollution. In scope of environmental engineering the coastal mussel populations and aquaculture appears to be a biological interface between society and the sea and they can be important tool of the coastal water management.

Great filtration capacity of seston-eating bivalves coupled with high efficiency of suspended matter retention principally determines mussels as a powerful natural biofilter which often plays the main role in coastal ecosystems. During the season of vegetation and growth some local mussel populations are able to filter out ones/twice the water volumes of the whole bays.

Mussels possess very high bioaccumulation capacity and resistance to pollution stress, thus the application of mussel aquaculture for environmental engineering is able to improve the recycling of nutrients from wastewater to bioproduction and greatly reduce the risk of environmental degradation in the coastal zone.

Mussel aquaculture as a relatively new approach of environmental engineering has been proposed for wastewater management and pollution control in the coastal waters.

References

[1] Gudimov A V 1993 *Arctic Seas: Bioindication of the Environment, Biotesting and Technology of Pollution Destruction* (Apatity: KSC RAS) 36–44

[2] Massabuau J-C, Gudimov A and Blanc P 2015 Environmental Monitoring of Arctic Waters with Unmanned Bivalve Biosensor Technology: One Year of Background Data Acquisition in the Barents Sea *Proc SPE Russian Petroleum Technology Conference* Paper SPE-176681-MS Available at: https://SPE176681MS?id=conferencepaper%2FSPE-176681-MS (accessed: 06.03.2016)

[3] Gudimov A V and Komarova E P 2021 Environmental control: from inertia of the standard biomonitoring to speed of the online one *IOP Conf. Series: Earth and Environ. Sc.* **625** 012013

[4] Winter J E 1969 Uber den Einfluss der Nahrungskonzentration und anderer Faktoren auf Filtrierleistung und Nahrungausnutzung der Muscheln Arctica islandica and Modiolus modiolus *Mar. Biol.* **4** 87–135

[5] Riisgard H U 2001 On measurement of filtration rates in bivalves—the stony road to reliable

IOP Conf. Series: Earth and Environmental Science **938** (2021) 012021 doi:10.1088/1755-1315/938/1/012021

data: review and interpretation *Mar. Ecol. Prog. Ser.* **211** 275–291

[6] Jørgensen C B 1949 The rate of feeding by Mytilus edulis in different kinds of suspension *JMBA* **28** 333–344

[7] Riisgard H U, Egede P P and Saavedra I B 2011 Feeding behaviour of the mussel, Mytilus edulis: new observations, with a minireview of current knowledge *J. Mar. Biol.* 312459

[8] Voskresenski K A 1948 *Proc. GOIN* **6(18)** 15–120

[9] *Marine mussels: their ecology and physiology* 1976 (Cambr. Univ. Press.) p 206

[10] Vahl O 1972 Particle retention and relation between water transport and oxygen uptake in Chlamys opercularis (L.) *Ophelia* **10** 67–74

[11] Riisgård H U, Kittner Ch and Seerup D F 2003 Regulation of opening state and filtration rate in filter-feeding bivalves (Cardium edule Mytilus edulis Mya arenaria) in response to low algal concentration *J. Exp. Mar. Biol. Ecol.* **284** 105–127

[12] Alimov A F 1981 *Functional ecology of freshwater bivalves* (Leningrad: Nauka) p 248

[13] Karnaukhov V N 1979 The role of filtrator molluscs rich in carotenoids in the self-cleaning of fresh waters *Proc. Symp. Biol. Hung.* **19** 151–167 (Budapest)

[14] Gudimov A V and et al 2002 Towards abilities of intertidal bivalves on nutrient recycling and restoration of the coastal environment in terms of physiological ecology *Abs. VIII Inter. Cong. Ecol.* 341(Seoul)

[15] Vanstralen M R and Dijkema R D 1994 Mussel culture in a changing environment - the effects of a coastal engineering project on mussel culture (Mytilus edulis L.) in the Oosterschelde estuary (the Netherlands) *Hydrobiologia* **283** 359–379

[16] Prins T S and Smaal A C 1994 *Hydrobiologia* **283** 413–429

[17] Gudimov A V 1989 *Feeding Relations of Bottom Invertebrates and Bottom Fishes of the Barents Sea* (Apatity: KSC RAS) p 96–103

[18] Alyakrinskaya I O 1966 *Zoological J.* **45** 998–1003

[19] Lee R F and et al 1972 Petroleum hydrocarbons: uptake and discharge by marine mussel, Mytilus edulis *Science* **177** 344–346

[20] Rantamaki P 1997 Release and retention of selected polycyclic aromatic hydrocarbons (PAH) and their methylated derivates by the common mussel (Mytilus edulis) in the brackish water of the Baltic Sea *Chemosphere* **35** 487–502

AUTHOR INDEX

Abakumov, E ... 107
Abdulkhalikov, Z.A. ... 66
Abramov, V A ... 86
Akhmedkhanova, Raisat ... 113
Akhmedov, M.E. ... 66
Alakayeva, Albina ... 113
Aliyeva, Samira ... 113
Andronov, E ... 107
Antoshina, T N ... 35
Boyakhchyan, A A ... 35
Burlov, V G ... 86
Bykovsky, N A ... 9
Cheng, Jinrui ... 1
Cherkasova, M ... 58
Danilova, A.A. ... 31
Demirova, A.F. ... 66
Dmitriev, S ... 58
Dyakov, A ... 144
Fanakova, N N ... 9
Galimov, D M ... 16
Galochkina, N A ... 136
Gladkov, G ... 107
Glotova, I A ... 136
Grigoriev, Victor S ... 22
Gudimov, AV ... 150
Hiramagomedova, Patimat ... 113
Ivakhnuk, G K ... 35
Kalashnik, A ... 144
Kantor, E A ... 9, 50
Kapitsyn, S Yu ... 86
Katraeva, Inna V ... 120
Khaidarov, A G ... 35
Khmeleva, O V ... 16
Khorin, B.V. ... 31
Kimeklis, A ... 107
Koroleva, L A ... 35
Korysheva, N N ... 136
Kovalev, Andrey A ... 22, 120
Kovalev, Dmitriy A ... 22, 120
Kuksov, M ... 58
Kutepov, Yu I ... 42
Kutepov, Yu Yi ... 42
Kutepova, N A ... 42
Labutina, N.D. ... 31
Litti, Yuriy V ... 22, 120
Liubarskaia, M A ... 100
Lykov, I N ... 93
Malkova, M A ... 9, 50
Manokhin, M. V. ... 73
Manokhin, V. Ya. ... 73
Mikhailov, G G ... 16
Mikhailova, N V ... 127
Minchenkov, N D ... 50
Mironova, M V ... 86
Mukhina, A S ... 42
Musayeva, Irina ... 113
Obolenskaya, E V ... 93
Parshina, Sofiya N ... 120
Puchkova, L N ... 9
Putinceva, N A ... 100
Qu, Jili ... 1
Rahman, P A ... 9
Rakhmanova, M.M. ... 66
Ryazanov, A G ... 16
Samukov, A ... 58
Savina, Y D ... 16
Senin, A V ... 16
Shabanov, Haji ... 113
Shakhov, S V ... 136
Smirnov, R D ... 42
Tikhonov, G S ... 136
Titov, S A ... 136
Vasileva, A D ... 42
Vasileva, A I ... 50
Vlasov, A.B. ... 31
Wang, Junfeng ... 1
Yasinskaya, A V ... 127
Yurin, D.A. ... 31
Yurina, N.A. ... 31
Zagirov, N.G. ... 66
Zhuravleva, Elena A ... 120
Zverev, A ... 107